Cambridge Elementary Classics

General Editor

J. H. SLEEMAN, M.A.
Professor of Classics in the University of London

MARTIAL AND PLINY

VESUVIUS

Martial and Pliny

Selections from

Martial and Pliny the Younger

Edited with Introduction, Notes and Vocabulary
by

E. C. KENNEDY, M.A.
Formerly Assistant Master at Malvern College

CAMBRIDGE UNIVERSITY PRESS

Published by the Syndics of the Cambridge University Press
Bentley House, 200 Euston Road, London NW1 2DB
American Branch: 32 East 57th Street, New York, N.Y. 10022

ISBN: 0 521 05683 7

First published 1942
Reprinted 1952 1960 1963 1966 1969 1974

Printed in Great Britain
at the University Printing House, Cambridge
(Brooke Crutchley, University Printer)

CONTENTS

PREFACE

This book is planned on exactly the same lines as my *Four Latin Authors*, though these extracts are slightly shorter (450 lines of Martial and 580 of Pliny) and the notes are rather fuller, as they must be when dealing with authors whose background is unfamiliar to the average boy or girl of School Certificate standard. It seems a pity that most pupils nowadays give up Latin without reading any of the post-Augustan writers, and probably without hearing of their existence at all. This edition of two of the major (and most interesting) Silver Latin authors is intended for the School Certificate year, and may provide a welcome change from Caesar and Ovid, without prejudicing the candidate's chances of doing well in the Unseen Translation; the notes and vocabulary bring them well within the capacity of the ordinary pupil, and the book may arouse interest by giving a view of daily life at Rome such as is never given in the authors usually read in the middle school.

The text of Martial is that of F. A. Paley and W. H. Stone, and of Pliny that of R. C. Kukula in the Teubner series. I have consulted editions of selections from Martial by Paley and Stone, and by H. M. Stephenson; and of selections from Pliny by

E. T. Merrill and by G. B. Allen; the translation of both authors in the Loeb series, and of Martial published anonymously by Messrs G. Bell and Sons, have also been useful. My thanks are due to Professor J. H. Sleeman for his many valuable suggestions.

I wish to take this opportunity of expressing my gratitude to my old friend and former Head Master, Mr R. R. Conway, from whose teaching and scholarship I first learned to appreciate the authors of the Silver Age of Latin Literature.

E. C. KENNEDY

Malvern
April 1942

DATES OF ROMAN EMPERORS

and of the chief events in the lives of MARTIAL and PLINY

Note. The term 'Silver Latin' is applied to the Post-Augustan literature of the Empire, when the language was beginning to decline from the standard of the Augustan or 'Golden' Age.

Augustus, 27 B.C.–A.D. 14.	
Tiberius, A.D. 14–37.	
Caligula, 37–41.	Martial born, about 40.
Claudius, 41–54.	
Nero, 54–68.	Pliny born, 61 or 62. Martial comes to Rome, 63 or 64.
Galba Otho } 68–69. Vitellius	
Vespasian, 69–79.	
Titus, 79–81.	Eruption of Vesuvius, 79. Pliny's legal career begins, 80.
Domitian, 81–96.	Martial's First Book of Epigrams published, 85 or 86. Pliny appointed *quaestor*, 89, and *praetor*, 93.
Nerva, 96–98.	Martial leaves Rome for Spain, 97 or 98.
Trajan, 98–117.	Pliny appointed consul, 100. Publication of his *Panegyric on Trajan.* Martial's Twelfth Book of Epigrams published about 102. Death of Martial, about 104. Pliny appointed Governor of Bithynia, 111. Death of Pliny, about 113.

HINTS ON SCANSION

The reason why you should be able to 'scan' a line, i.e. mark the long and short syllables, is partly to be able to read the verse correctly and know what the poet is doing, and partly because some words, e.g. the nominative and ablative singular of the 1st declension, can sometimes be distinguished only by the metre. A line of Latin poetry is divided into a certain number of 'feet', which contain a certain combination of syllables. Martial's poems are mostly 'Elegiac couplets', in which the first longer line is a Hexameter, containing six 'feet', which are either 'dactyls' (– ∪ ∪) or 'spondees' (– –), and the second shorter line is a Pentameter of five 'feet', of which one is a spondee made up of two half-spondees, i.e. one long syllable ending each half of the line. The scheme is

1. – ∪ ∪ | – ∪ ∪ | – ∪ ∪ | – ∪ ∪ | – ∪ ∪ | – ⏓
   – – | – – | – – | – –

Hexameter.

2. – ∪ ∪ | – ∪ ∪ | – || – ∪ ∪ | – ∪ ∪ | ⏓
   – – | – –

Pentameter.

You should learn the following six easy rules (there are more, of course, but these are the easiest and most important).

1. A word ending in a vowel or *-m* which is followed by a word beginning with a vowel (or *h*) has its last syllable cut off or 'elided', so that it does not count at all in scansion.

2. All diphthongs, i.e. combinations of two vowels, like the *ae* of *mensae*, are long (*qu* does not count).

3. A vowel before two consonants, either in the same word or with one at the end of one word and another at the beginning of the next, forms a long syllable. If *r* or *l* is the second consonant (in the same word) the syllable can sometimes be either short or long, e.g. *patrem* can have its first syllable either short or long. *x* and *z* count as double consonants, and *h* does not count at all.

4. A vowel before another vowel in the same word is usually short, e.g. *rĕī*. The *i* in, for example, *iam* and *iaceo* is a consonant, pronounced *y*.

5. The final *-a* of the 1st declension ablative singular is long; most other final *-a*'s are short.

6. Final *-i* and *-o* are usually long, but in Silver Latin the final *-o* is often made short.

When you start scanning a Hexameter, first of all look for any elisions according to rule 1. Then mark off the last five syllables, which are always – ◡ ◡ | – ⏓, like 'bláckberry púdding'. Next, count up the remaining syllables; if there are 12, there will be 4 dactyls, if 11, 3 dactyls and 1 spondee, if 10, 2 dactyls and 2 spondees, if 9, 3 spondees and 1 dactyl, and if 8, 4 spondees. Then mark off all the syllables which you know according to the other rules, remembering that a syllable between two longs must itself be long. There will be some which you do not know for certain; but your pronunciation of the line may help. You ought also to mark the 'caesura', or break between two words in the 3rd or 4th foot, with a thin wavy line.

The Pentameter is much easier to scan, as the last 8 syllables are always the same: – ‖ – ⏑ ⏑ | – ⏑ ⏑ | ⏓. The last 7 syllables sound like the English 'bláckberry ráspberry píe', and can be marked off at once. There is a break in the middle of the line just before this, to be marked with two downward strokes, and the syllable before this is always long. This leaves only the first two feet, which will contain either 6 syllables (i.e. 2 dactyls), 5 (1 dactyl and 1 spondee), or 4 (2 spondees), so you will have little difficulty in scanning them.

Martial also uses Hendecasyllables, or lines of 11 syllables, as follows: – – | – ⏑ ⏑ | – ⏑ | – ⏑ | – ⏓. The foot – ⏑ is called a 'trochee'. This metre is very pleasant to read and also easy to scan, as each line is just the same as all the others, and no variations are allowed. Tennyson wrote some Hendecasyllables, of which two lines are:

> Lóok, I cóme to the tést, a tíny póem
> Áll compósed in a métre óf Catúllus.

Martial uses a metre called 'scazons' or 'limping lines' in XXVII. This is based on the iambic line of six feet called 'iambi', ⏑ –, but in scazons the sixth foot is always a spondee, which makes the line halt or 'limp'. The first and third feet can also be spondees, and in any of the first four feet the long syllable can be 'resolved' into two short syllables, though in this particular epigram Martial does this only three times (ll. 22, 24, 29), each time in the first foot, where the spondee becomes an 'anapaest', ⏑ ⏑ –. The scheme therefore in this poem is

⏑ –	⏑ –	⏑ –	⏑ –	⏑ –	– ⏓
– –		– –			
⏑ ⏑ –					

Notice that in all lines, hexameters, pentameters, hendecasyllables and scazons, the last syllable is sometimes short instead of long.

MARTIAL

Marcus Valerius Martialis was born about A.D. 40 at Bilbilis, a Roman colony on the river Salo in Spain; his parents were probably Spaniards who had received Roman citizenship. He was well educated and came to Rome in A.D. 63 or 64, where he lived perhaps as the client of his fellow-countrymen the Senecas until their ruin in the conspiracy of Piso in 65. He then had to depend on the patronage of some other great family, and among others Pliny the Younger helped to support him later on. His first work was the *Liber Spectaculorum*, poems on the games with which Titus opened the Colosseum in A.D. 80, followed by a series of couplets to accompany presents sent to friends. In 85 or 86 came his first book of epigrams, which at once established his fame, and other books were published at intervals of about a year, until by 96 he had produced eleven. Book XII was written in about 102, after his return to Spain in 97 or 98. He lived at Rome for thirty-four years, and made many friends in literary circles, such as Lucan, Juvenal, Pliny, Quintilian and Silius Italicus. His reputation as a poet was high, and though Domitian refused to give him money he regarded Martial with favour and confirmed the *ius trium liberorum*, the privileges of a father of three children, which Titus had given him previously—a proof that he was known at court as early as 81. He was also given a military tribunate, which carried with it equestrian privileges, and no doubt many rich men were glad to count him among their clients and to give him the daily dole, *sportula*, besides other presents and invitations to dinner. He speaks often of his poverty, but he possessed slaves, a carriage and pair, and eventually a house of his own and a small villa at Nomentum. His life at Rome was the usual social round, often described in his poems, which he says interfered with his literary work. In 97 or 98 he grew tired of city life and returned to Spain, where he lived on an estate given him by a Spanish lady called Marcella. It is probable that he was unmarried, and that he was most likely not speaking for himself in poems in which he refers to a wife. He died in Spain some time after the publication of his last book in about 102. His friend and patron Pliny, in a letter (III, 21) generally assigned to 104, refers to his death as having recently occurred.

The epigram in Greek literature was at first an inscription on a monument or tombstone, but this developed into a short poem on any subject, person or incident, a type of writing which Catullus introduced to Latin, but which Martial developed more skilfully than any other poet. It generally consists of an introduction or description, followed by the 'point' which the poet makes, usually in the last line. Some of Martial's epigrams are genuine epitaphs, others are brief tales in verse, others letters addressed to the poet's patrons asking for assistance, or descriptions, usually satirical, of contemporary characters or events; many bad characters are held up to ridicule. The metres used are mainly the elegiac couplet and hendecasyllables; a description of the chief metres is given on pp. xi–xiv. The greatest objection against Martial's character is his fulsome flattery of the emperor Domitian, but he himself says that he had to obtain court patronage to secure a livelihood, and this was the only way in which a poet could do so; some of his poems are also unpleasant in tone, but that was a tendency common to the age. As an epigrammatic poet Martial stands alone. His verse is vigorous, polished, and of infinite variety, expressing a love of nature and also exquisite pathos as well as biting satire and keen wit. In addition he paints the everyday life of Rome in vivid colours and makes it seem very near our own times.

The Relative Social Positions of Martial and Pliny

It should be emphasized that though Martial and Pliny were both men of letters, living at about the same period—Martial was roughly twenty years older than Pliny—and acquainted with one another, they occupied very different positions in society. Pliny was an ex-consul and was held in high esteem at court, and was also a man of great wealth, whereas Martial held no official post and was always in need of money; he was constantly asking his rich friends for money and other gifts, which no doubt he often received, though not often enough to keep him comfortably in the fashionable style of the day.

At Rome in the first century A.D. the honourable relationship between *patronus* and *cliens* which was a feature of the Republic had changed into something quite different. The patron was surrounded by a large number of clients, who made it their profession to attach themselves to one or more of the rich men of the time. It was their duty to attend the *salutatio* or morning levée of their patron, to accompany him on his social or official business, and generally to accommodate themselves to his wishes; in return for this they used at one time to be entertained to dinner, but the increase in the numbers of clients made this impossible, and they received instead the *sportula*, originally a basket of food, for which later a sum of money (100 *quadrantes* or 25 *asses*, about 1*s*. 6*d*.) was substituted; this was a daily payment, no doubt in some cases an important addition to the client's income. Martial as a popular poet was of course much better off than the average client, but he was always dependent on the good-will of his patrons, among whom he numbered Pliny. Pliny thought highly of Martial's powers, and gave him a sum of money for his journey to Spain. In the letter included in this volume in which he records the poet's death (Pliny, III) he speaks of him in very friendly terms and quotes part of a flattering epigram addressed to him by Martial. He makes an interesting and appreciative estimate of Martial's character, but obviously thinks that his own literary work will survive and that Martial's is doomed to perish.

MARTIAL

I. (Book I, 1)

To his readers, introducing himself.
Hendecasyllables (see p. xiii).

Hic est quem legis ille, quem requiris,
Toto notus in orbe Martialis
Argutis epigrammaton libellis:
Cui, lector studiose, quod dedisti
Viventi decus atque sentienti,
Rari post cineres habent poetae.

II. (I, 3)

To his book, advising it not to be too eager to be published.
Elegiacs (see p. xi).

Argiletanas mavis habitare tabernas,
  Cum tibi, parve liber, scrinia nostra vacent.
Nescis, heu, nescis dominae fastidia Romae:
  Crede mihi, nimium Martia turba sapit.
Maiores nusquam rhonchi, iuvenesque senesque
  Et pueri nasum rhinocerotis habent.
Audieris cum grande sophos, dum basia iactas,
  Ibis ab excusso missus in astra sago.
Sed tu ne totiens domini patiare lituras
  Neve notet lusus tristis harundo tuos,
Aetherias, lascive, cupis volitare per auras:
  I, fuge; sed poteras tutior esse domi.

III. (I, 4)

To the emperor Domitian, asking him not to take his book too seriously. Elegiacs.

Contigeris nostros, Caesar, si forte libellos,
Terrarum dominum pone supercilium.
Consuevere iocos vestri quoque ferre triumphi,
Materiam dictis nec pudet esse ducem.
Qua Thymelen spectas derisoremque Latinum,
Illa fronte precor carmina nostra legas.
Innocuos censura potest permittere lusus:
Lasciva est nobis pagina, vita proba.

IV. (I, 13)

On the noble death of Arria. Elegiacs.

Casta suo gladium cum traderet Arria Paeto,
Quem de visceribus strinxerat ipsa suis,
'Si qua fides, vulnus quod feci non dolet,' inquit;
'Sed quod tu facies, hoc mihi, Paete, dolet.'

V. (I, 70)

To his book, telling it to go to the house of Julius Proculus, and to make excuses for Martial's not calling on him in person. Elegiacs.

Vade salutatum pro me, liber: ire iuberis
Ad Proculi nitidos, officiose, lares.
Quaeris iter, dicam: vicinum Castora canae
Transibis Vestae virgineamque domum;
Inde sacro veneranda petes Palatia clivo,
Plurima qua summi fulget imago ducis.

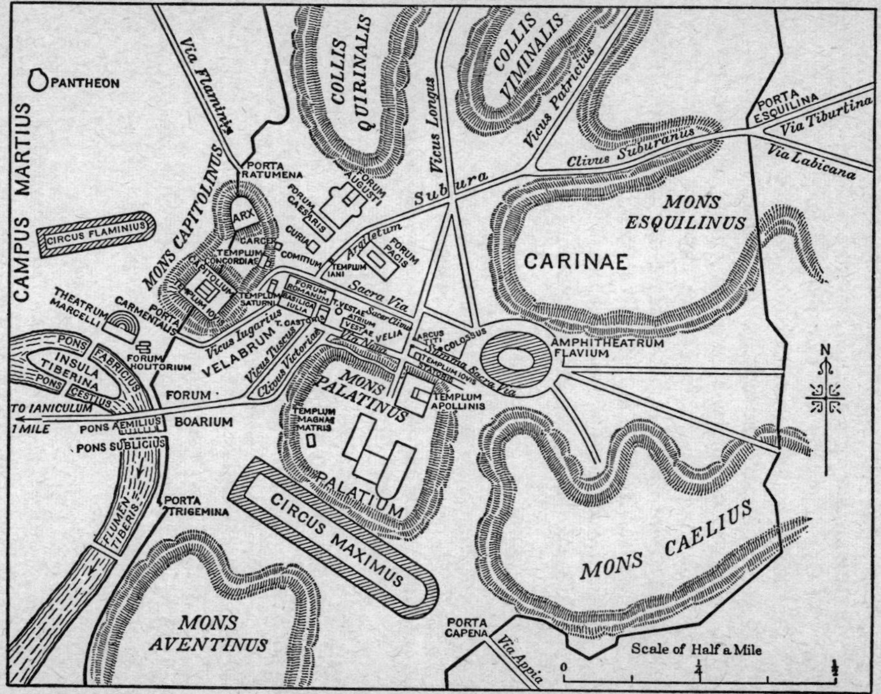

PLAN OF CENTRAL ROME ABOUT A.D. 90

Nec te detineat miri radiata colossi
Quae Rhodium moles vincere gaudet opus.
Flecte vias hac qua madidi sunt tecta Lyaei
Et Cybeles picto stat Corybante tholus.
Protinus a laeva clari tibi fronte Penates
Atriaque excelsae sunt adeunda domus.
Hanc pete, nec metuas fastus limenque superbum:
Nulla magis toto ianua poste patet,
Nec propior quam Phoebus amet doctaeque sorores.
Si dicet 'Quare non tamen ipse venit?'
Sic licet excuses 'Quia qualiacunque leguntur
Ista, salutator scribere non potuit.'

VI. (I, 88)

Epitaph on Alcimus, a favourite slave of Martial's, who died young. Elegiacs.

Alcime, quem raptum domino crescentibus annis
Labicana levi caespite velat humus,
Accipe non Pario nutantia pondera saxo,
Quae cineri vanus dat ruitura labor,
Sed faciles buxos et opacas palmitis umbras
Quaeque virent lacrimis roscida prata meis.
Accipe, care puer, nostri monimenta doloris:
Hic tibi perpetuo tempore vivet honor.
Cum mihi supremos Lachesis perneverit annos,
Non aliter cineres mando iacere meos.

VII. (I, 109)

On a pet dog called Issa. Hendecasyllables.

Issa est passere nequior Catulli,
Issa est purior osculo columbae,
Issa est blandior omnibus puellis,

Issa est carior Indicis lapillis,
Issa est deliciae catella Publi.
Hanc tu, si queritur, loqui putabis;
Sentit tristitiamque gaudiumque.
Collo nixa cubat capitque somnos,
Ut suspiria nulla sentiantur.
Hanc ne lux rapiat suprema totam,
Picta Publius exprimit tabella,
In qua tam similem videbis Issam,
Ut sit tam similis sibi nec ipsa.
Issam denique pone cum tabella:
Aut utramque putabis esse veram,
Aut utramque putabis esse pictam.

'Beware of the Dog'. A floor-mosaic from the entrance of a house at Pompeii.

VIII. (I, 117)

To Lupercus, a mean acquaintance, who wanted to borrow a copy of Martial's poems instead of buying one. Hendecasyllables.

Occurris quotiens, Luperce, nobis,
'Vis mittam puerum' subinde dicis,
'Cui tradas epigrammaton libellum,
Lectum quem tibi protinus remittam?'
Non est quod puerum, Luperce, vexes.
Longum est, si velit ad Pirum venire,
Et scalis habito tribus, sed altis.
Quod quaeris propius petas licebit.
Argi nempe soles subire Letum:
Contra Caesaris est forum taberna
Scriptis postibus hinc et inde totis,
Omnes ut cito perlegas poetas.
Illinc me pete, nec roges Atrectum,—
Hoc nomen dominus gerit tabernae,—
De primo dabit alterove nido
Rasum pumice purpuraque cultum
Denaris tibi quinque Martialem.
'Tanti non es' ais? Sapis, Luperce.

IX. (II, 75)

On a tame lion which killed two boys in the arena. Elegiacs.

Verbera securi solitus leo ferre magistri
  Insertamque pati blandus in ora manum
Dedidicit pacem subito feritate reversa,
  Quanta nec in Libycis debuit esse iugis.

Nam duo de tenera puerilia corpora turba,
Sanguineam rastris quae renovabat humum,
Saevus et infelix furiali dente peremit:
Martia non vidit maius harena nefas.
Exclamare libet: 'crudelis, perfide, praedo,
A nostra pueris parcere disce lupa!'

X. (II, 87)

To a lady-killer. Elegiacs.

Dicis amore tui bellas ardere puellas,
Qui faciem sub aqua, Sexte, natantis habes.

XI. (III, 5)

To his book, telling it to go to the home of Julius Cerealis, to whom it would need no introduction. Elegiacs.

Vis commendari sine me cursurus in urbem,
Parve liber, multis, an satis unus erit?
Unus erit, mihi crede, satis, cui non eris hospes,
Iulius, adsiduum nomen in ore meo.
Protinus hunc primae quaeres in limine Tectae:
Quos tenuit Daphnis, nunc tenet ille lares.
Est illi coniunx, quae te manibusque sinuque
Excipiet, tu vel pulverulentus eas.
Hos tu seu pariter sive hanc illumve priorem
Videris, hoc dices 'Marcus havere iubet,'
Et satis est: alios commendet epistula: peccat
Qui commendandum se putat esse suis.

XII. (III, 19)

On a boy who was killed by a viper hidden in the mouth of a bronze bear. Elegiacs.

Proxima centenis ostenditur ursa columnis,
  Exornant fictae qua platanona ferae.
Huius dum patulos adludens temptat hiatus
  Pulcher Hylas, teneram mersit in ora manum.
Vipera sed caeco scelerata latebat in aere
  Vivebatque anima deteriore fera.
Non sensit puer esse dolos, nisi dente recepto,
  Dum perit: o facinus, falsa quod ursa fuit!

XIII. (III, 38)

To Sextus, telling him how difficult it was for an honest man to live at Rome except by chance. Elegiacs.

Quae te causa trahit vel quae fiducia Romam,
  Sexte? quid aut speras aut petis inde? refer.
'Causas' inquis 'agam Cicerone disertior ipso
  Atque erit in triplici par mihi nemo foro.'
Egit Atestinus causas et Civis (utrumque
  Noras); sed neutri pensio tota fuit.
'Si nihil hinc veniet, pangentur carmina nobis:
  Audieris, dices esse Maronis opus.'
Insanis: omnes gelidis quicunque lacernis
  Sunt ibi, Nasones Vergiliosque vides.
'Atria magna colam.' Vix tres aut quattuor ista
  Res aluit, pallet cetera turba fame.
'Quid faciam? suade: nam certum est vivere Romae.'
  Si bonus es, casu vivere, Sexte, potes.

XIV. (IV, 8)

To Euphemus, Domitian's table-steward, describing the daily round of life at Rome, and asking him to present Martial's book to the emperor after dinner. Elegiacs.

Prima salutantes atque altera conterit hora,
Exercet raucos tertia causidicos:
In quintam varios extendit Roma labores,
Sexta quies lassis, septima finis erit:
Sufficit in nonam nitidis octava palaestris,
Imperat extructos frangere nona toros:
Hora libellorum decima est, Eupheme, meorum,
Temperat ambrosias cum tua cura dapes,
Et bonus aetherio laxatur nectare Caesar
Ingentique tenet pocula parca manu.
Tunc admitte iocos: gressu timet ire licenti
Ad matutinum nostra Thalia Iovem.

XV. (IV, 14)

To Silius Italicus, the poet, asking him to unbend during the Saturnalia and read Martial's sportive verses. Hendecasyllables.

Sili, Castalidum decus sororum,
Qui periuria barbari furoris
Ingenti premis ore perfidosque
Astus Hannibalis levesque Poenos
Magnis cedere cogis Africanis:
Paulum seposita severitate,
Dum blanda vagus alea December
Incertis sonat hinc et hinc fritillis
Et ludit tropa nequiore talo,
Nostris otia commoda Camenis.

Nec torva lege fronte, sed remissa
Lascivis madidos iocis libellos.
Sic forsan tener ausus est Catullus
Magno mittere passerem Maroni.

XVI. (IV, 30)

To a fisherman at Baiae, warning him not to catch fish from the emperor's pool. Hendecasyllables.

Baiano procul a lacu recede,
Piscator; fuge, ne nocens recedas.
Sacris piscibus hae natantur undae,
Qui norunt dominum manumque lambunt
Illam, qua nihil est in orbe maius.
Quid, quod nomen habent et ad magistri
Vocem quisque sui venit citatus?
Hoc quondam Libys impius profundo,
Dum praedam calamo tremente ducit,
Raptis luminibus repente caecus
Captum non potuit videre piscem,
Et nunc sacrilegos perosus hamos
Baianos sedet ad lacus rogator.
At tu, dum potes, innocens recede
Iactis simplicibus cibis in undas,
Et pisces venerare delicatos.

XVII. (IV, 64)

A description of the charming suburban villa of the poet's friend Julius Martialis. Hendecasyllables.

Iuli iugera pauca Martialis
Hortis Hesperidum beatiora
Longo Ianiculi iugo recumbunt:

Reconstruction of an *insula*, or block of flats, at Ostia.
See Martial VIII, 7, and XXXI, 2.

Lati collibus imminent recessus
Et planus modico tumore vertex
Caelo perfruitur sereniore,
Et curvas nebula tegente valles
Solus luce nitet peculiari:
Puris leniter admoventur astris
Celsae culmina delicata villae.
Hinc septem dominos videre montes
Et totam licet aestimare Romam,
Albanos quoque Tusculosque colles
Et quodcunque iacet sub urbe frigus,
Fidenas veteres brevesque Rubras,
Et quod virgineo cruore gaudet
Annae pomiferum nemus Perennae.
Illinc Flaminiae Salariaeque

Gestator patet essedo tacente,
Ne blando rota sit molesta somno,
Quem nec rumpere nauticum celeuma,
Nec clamor valet helciariorum,
Cum sit tam prope Mulvius, sacrumque
Lapsae per Tiberim volent carinae.
Hoc rus, seu potius domus vocanda est,
Commendat dominus: tuam putabis;
Tam non invida tamque liberalis,
Tam comi patet hospitalitate.
Credas Alcinoi pios Penates,
Aut facti modo divitis Molorchi.
Vos nunc omnia parva qui putatis,
Centeno gelidum ligone Tibur
Vel Praeneste domate pendulamque
Uni dedite Setiam colono:
Dum me iudice praeferantur istis
Iuli iugera pauca Martialis.

XVIII. (V, 34)

On the death of a little slave girl called Erotion. Elegiacs.

Hanc tibi, Fronto pater, genetrix Flaccilla, puellam
  Oscula commendo deliciasque meas,
Parvula ne nigras horrescat Erotion umbras
  Oraque Tartarei prodigiosa canis.
Inpletura fuit sextae modo frigora brumae,
  Vixisset totidem ni minus illa dies.
Inter tam veteres ludat lasciva patronos
  Et nomen blaeso garriat ore meum.
Mollia non rigidus caespes tegat ossa, nec illi,
  Terra, gravis fueris: non fuit illa tibi.

XIX. (VII, 19)

On a fragment of the ship Argo. Elegiacs.

Fragmentum quod vile putas et inutile lignum,
Haec fuit ignoti prima carina maris.
Quam nec Cyaneae quondam potuere ruinae
Frangere nec Scythici tristior ira freti,
Saecula vicerunt: sed quamvis cesserit annis,
Sanctior est salva parva tabella rate.

XX. (VII, 61)

To Domitian, congratulating him on keeping the streets of Rome clear of the encroachments of shop-keepers. Elegiacs.

Abstulerat totam temerarius institor urbem
Inque suo nullum limine limen erat.
Iussisti tenues, Germanice, crescere vicos,
Et modo quae fuerat semita, facta via est.
Nulla catenatis pila est praecincta lagonis,
Nec praetor medio cogitur ire luto;
Stringitur in densa nec caeca novacula turba,
Occupat aut totas nigra popina vias.
Tonsor, copo, cocus, lanius sua limina servant.
Nunc Roma est, nuper magna taberna fuit.

XXI. (VII, 96)

Epitaph on Urbicus, who died in infancy. Elegiacs.

Conditus hic ego sum Bassi dolor, Urbicus infans,
Cui genus et nomen maxima Roma dedit.
Sex mihi de prima derant trieteride menses,
Ruperunt tetricae cum male pensa deae.

Quid species, quid lingua mihi, quid profuit aetas?
Da lacrimas tumulo, qui legis ista, meo.
Sic ad Lethaeas, nisi Nestore serius, undas
Non eat, optabis quem superesse tibi.

XXII. (VIII, 69)

To Vacerra, a critic who praised only dead poets. Hendecasyllables.

Miraris veteres, Vacerra, solos,
Nec laudas nisi mortuos poetas.
Ignoscas petimus, Vacerra: tanti
Non est, ut placeam tibi, perire.

XXIII. (IX, 61)

On a plane-tree in Corduba, which was planted by the hand of Julius Caesar. Elegiacs.

In Tartessiacis domus est notissima terris,
Qua dives placidum Corduba Baetin amat,
Vellera nativo pallent ubi flava metallo
Et linit Hesperium bractea viva pecus;
Aedibus in mediis totos amplexa penates
Stat platanus densis Caesariana comis,
Hospitis invicti posuit quam dextera felix,
Coepit et ex illa crescere virga manu.
Auctorem dominumque suum sentire videtur:
Sic viret et ramis sidera celsa petit.
Saepe sub hac madidi luserunt arbore Fauni,
Terruit et tacitam fistula sera domum;
Dumque fugit solos nocturnum Pana per agros,
Saepe sub hac latuit rustica fronde Dryas.

Atque oluere lares comissatore Lyaeo,
Crevit et effuso laetior umbra mero;
Hesternisque rubens deiecta est herba coronis,
Atque suas potuit dicere nemo rosas.
O dilecta deis, o magni Caesaris arbor,
Ne metuas ferrum sacrilegosque focos.
Perpetuos sperare licet tibi frondis honores:
Non Pompeianae te posuere manus.

XXIV. (IX, 68)

To a schoolmaster, who woke his neighbours up by his noisy lessons in the early morning. Elegiacs.

Quid tibi nobiscum est, ludi scelerate magister,
Invisum pueris virginibusque caput?
Nondum cristati rupere silentia galli:
Murmure iam saevo verberibusque tonas.
Tam grave percussis incudibus aera resultant,
Causidicum medio cum faber aptat equo:
Mitior in magno clamor furit amphitheatro,
Vincenti parmae cum sua turba favet.
Vicini somnum non tota nocte rogamus:
Nam vigilare leve est, pervigilare grave est.
Discipulos dimitte tuos. Vis, garrule, quantum
Accipis ut clames, accipere ut taceas?

XXV. (IX, 70)

To Caecilianus, a profligate who complained of the morals of the age. Elegiacs.

Dixerat 'o mores! o tempora!' Tullius olim,
Sacrilegum strueret cum Catilina nefas,
Cum gener atque socer diris concurreret armis
Maestaque civili caede maderet humus.

Cur nunc 'o mores!' cur nunc 'o tempora!' dicis?
Quod tibi non placeat, Caeciliane, quid est?
Nulla ducum feritas, nulla est insania ferri;
Pace frui certa laetitiaque licet.
Non nostri faciunt, tibi quod tua tempora sordent,
Sed faciunt mores, Caeciliane, tui.

XXVI. (IX, 101)

A contrast between the labours of Hercules and the exploits of Domitian, who had recently built a temple for Hercules on the Appian Way. Elegiacs.

Appia, quam simili venerandus in Hercule Caesar
Consecrat, Ausoniae maxima fama viae,
Si cupis Alcidae cognoscere facta prioris,
Disce: Libyn domuit, aurea poma tulit;
Peltatam Scythico discinxit Amazona nodo,
Addidit Arcadio terga leonis apro;
Aeripedem silvis cervum, Stymphalidas astris
Abstulit, a Stygia cum cane venit aqua;
Fecundam vetuit reparari mortibus hydram,
Hesperias Tusco lavit in amne boves.
Haec minor Alcides: maior quae gesserit, audi,
Sextus ab Albana quem colit arce lapis.
Adseruit possessa malis Palatia regnis:
Prima suo gessit pro Iove bella puer.
Solus Iuleas cum iam retineret habenas,
Tradidit inque suo tertius orbe fuit.
Cornua Sarmatici ter perfida contudit Histri,
Sudantem Getica ter nive lavit equum.
Saepe recusatos parcus duxisse triumphos
Victor Hyperboreo nomen ab orbe tulit.

Templa deis, mores populo dedit, otia ferro,
  Astra suis, caelo sidera, serta Iovi.
Herculeum tantis numen non sufficit actis:
  Tarpeio deus hic commodet ora patri.

XXVII. (X, 30)

A description of the sea-side villa of Martial's friend Apollinaris at Formiae. Scazons (see p. xiii).

O temperatae dulce Formiae litus,
Vos, cum severi fugit oppidum Martis
Et inquietas fessus exuit curas,
Apollinaris omnibus locis praefert.
Non ille sanctae dulce Tibur uxoris,
Nec Tusculanos Algidosve secessus,
Praeneste nec sic Antiumque miratur.
Non blanda Circe Dardanisve Caieta
Desiderantur, nec Marica nec Liris,
Nec in Lucrina lota Salmacis vena.
Hic summa leni stringitur Thetis vento;
Nec languet aequor, viva sed quies ponti
Pictam phaselon adiuvante fert aura,
Sicut puellae non amantis aestatem
Mota salubre purpura venit frigus.
Nec saeta longo quaerit in mari praedam,
Sed e cubiclo lectuloque iactatam
Spectatus alte lineam trahit piscis.
Si quando Nereus sentit Aeoli regnum,
Ridet procellas tuta de suo mensa.
Piscina rhombum pascit et lupos vernas,
Natat ad magistrum delicata muraena;
Nomenculator mugilem citat notum

Et adesse iussi prodeunt senes mulli.
Frui sed istis quando, Roma, permittis?
Quot Formianos inputat dies annus
Negotiosis rebus urbis haerenti?
O ianitores vilicique felices!
Dominis parantur ista, serviunt vobis.

XXVIII. (X, 47)

To Julius Martialis, giving a list of the requirements for a happy life. Hendecasyllables.

Vitam quae faciunt beatiorem,
Iucundissime Martialis, haec sunt:
Res non parta labore, sed relicta;
Non ingratus ager, focus perennis;
Lis numquam, toga rara, mens quieta;
Vires ingenuae, salubre corpus;
Prudens simplicitas, pares amici;
Convictus facilis, sine arte mensa;
Nox non ebria, sed soluta curis;
Somnus, qui faciat breves tenebras:
Quod sis, esse velis nihilque malis;
Summum nec metuas diem, nec optes.

XXIX. (X, 103)

To his fellow-townsmen at Bilbilis, hoping for a friendly welcome after thirty-four years' absence in Rome. Elegiacs.

Municipes, Augusta mihi quos Bilbilis acri
  Monte creat, rapidis quem Salo cingit aquis,
Ecquid laeta iuvat vestri vos gloria vatis?
  Nam decus et nomen famaque vestra sumus,

Nec sua plus debet tenui Verona Catullo
Meque velit dici non minus illa suum.
Quattuor accessit tricesima messibus aestas,
Ut sine me Cereri rustica liba datis.
Moenia dum colimus dominae pulcherrima Romae,
Mutavere meas Itala regna comas.
Excipitis placida reducem si mente, venimus;
Aspera si geritis corda, redire licet.

XXX. (X, 104)

To his book, telling it to go with his friend Flavus to Bilbilis, and to remind Flavus to buy him a country retreat there. Hendecasyllables.

I nostro comes, i libelle, Flavo
Longum per mare, sed faventis undae,
Et cursu facili tuisque ventis
Hispanae pete Tarraconis arces.
Illinc te rota tollet et citatus
Altam Bilbilin et tuum Salonem
Quinto forsitan essedo videbis.
Quid mandem tibi, quaeris? Ut sodales
Paucos, sed veteres et ante brumas
Triginta mihi quattuorque visos
Ipsa protinus a via salutes
Et nostrum admoneas subinde Flavum,
Iucundos mihi nec laboriosos
Secessus pretio paret salubri,
Qui pigrum faciant tuum parentem.
Haec sunt. Iam tumidus vocat magister
Castigatque moras, et aura portum
Laxavit melior; vale, libelle:
Navem, scis puto, non moratur unus.

XXXI. (XI, 18)

To Lupus, who had given Martial a tiny country estate.
Hendecasyllables.

Donasti, Lupe, rus sub urbe nobis,
Sed rus est mihi maius in fenestra.
Rus hoc dicere, rus potes vocare?
In quo ruta facit nemus Dianae,
Argutae tegit ala quod cicadae,
Quod formica die comedit uno,
Clausae cui folium rosae corona est;
In quo non magis invenitur herba,
Quam Cosmi folium piperve crudum;
In quo nec cucumis iacere rectus,
Nec serpens habitare tota possit,
Urucam male pascit hortus unam,
Consumpto moritur culex salicto,
Et talpa est mihi fossor atque arator.
Non boletus hiare, non mariscae
Ridere aut violae patere possunt
Fines mus populatur et colono
Tamquam sus Calydonius timetur,
Et sublata volantis ungue Procnes
In nido seges est hirundinino;
22 Non est dimidio locus Priapo.
Vix inplet cochleam peracta messis
Et mustum nuce condimus picata.
Errasti, Lupe, littera sed una.
Nam quo tempore praedium dedisti,
Mallem tu mihi prandium dedisses.

XXXII. (XI, 53)

On Claudia Rufina, the British wife of Pudens, wishing her happiness. Elegiacs.

Claudia caeruleis cum sit Rufina Britannis
Edita, quam Latiae pectora gentis habet!
Quale decus formae! Romanam credere matres
Italides possunt, Atthides esse suam.
Di bene, quod sancto peperit fecunda marito,
Quod sperat generos quodque puella nurus.
Sic placeat superis, ut coniuge gaudeat uno
Et semper natis gaudeat illa tribus.

XXXIII. (XII, 18)

To the poet Juvenal, telling him how Martial had abandoned the heat and noise of Rome and was living a delightful country life at Bilbilis. Hendecasyllables.

Dum tu forsitan inquietus erras
Clamosa, Iuvenalis, in Subura,
Aut collem dominae teris Dianae;
Dum per limina te potentiorum
Sudatrix toga ventilat vagumque
Maior Caelius et minor fatigant:
Me multos repetita post Decembres
Accepit mea rusticumque fecit
Auro Bilbilis et superba ferro.
Hic pigri colimus labore dulci
Boterdum Plateamque; Celtiberis
Haec sunt nomina crassiora terris.
Ingenti fruor improboque somno,
Quem nec tertia saepe rumpit hora,

15 Et totum mihi nunc repono, quidquid
Ter denos vigilaveram per annos.
Ignota est toga, sed datur petenti
Rupta proxima vestis a cathedra.
Surgentem focus excipit superba
Vicini strue cultus iliceti,
21 Multa vilica quem coronat olla.
24 Dispensat pueris rogatque longos
Levis ponere vilicus capillos.
Sic me vivere, sic iuvat perire.

NOTES

MARTIAL

I. (Book I, 1)

1. Hic est ille...] 'here is the man whose poems (lit. whom) you are reading, whom you are asking for—Martial, known throughout the whole world for his witty books of epigrams'. *epigrammaton* is Greek genitive plural.

4. cui...quod...] 'the fame which you have given to him while still alive and conscious (*sentienti*) few poets obtain (even) after (they are) ashes'. The order of words is difficult, with the two relative pronouns; the antecedent of *quod*, *decus*, is inside the relative clause and is the object of *habent*.

II. (I, 3)

1. The *Argiletum*, stretching from the north side of the *Forum Romanum* to the *Subura*, was the district in which booksellers had their shops. See VIII, 9.

2. Cum] 'although'.

scrinia were circular boxes for holding the papyrus rolls which served as books. *nostra* = *mea*.

3. dominae...Romae] 'of imperial Rome'. *dominae* is here an adjective.

4. nimium sapit] 'is too critical', i.e. wants better verses than yours.

Martia turba] Mars was the father of Romulus, the founder of Rome, so *Martia* here = *Romana*.

5. nusquam] Supply *sunt*: 'nowhere are there...'.

-que...-que] 'both...and'.

6. Et pueri] 'and even boys'. To turn up the nose, like a rhinoceros, was (and still is) a sign of contempt.

7. Audieris (= *audiveris*) is future perfect: 'When you have heard a loud "Bravo!" and while you are throwing kisses (to your admirers), you will be flung (*ibis*) up towards the stars, thrown up from a blanket suddenly pulled tight', lit. shaken out.

sophōs is a Greek adverb meaning 'cleverly', here used as a

neuter noun. Tossing in a blanket (*sagum* is also a military cloak) was a trick common among soldiers.

9. **ne...patiare** (=*patiaris*): 'so that you may not endure your author's corrections, and that his stern pen may not mark your sportive verses (*lusus*)...'. The final -*o* of *harundo* is short, as final -*o* in the nominative often is in Silver Latin.

11. **volitare**] Like a bird leaving its nest.

12. **poteras**] 'you might have been safer at home', where no one would criticize you.

III. (I, 4)

1. **Contigeris**] Future perfect: 'if you happen to come across my (*nostros* often = *meos*) little books'.

2. 'Lay aside that frown which rules the world', lit. master of the earth.

3. **Consuevere** (=*consueverunt*): 'even the triumphs of you emperors (*vestri*, not *tui*, as he is addressing all the emperors) have become accustomed to endure jokes, and a general is not ashamed to be a subject for wit (*dictis*)'. Soldiers used to make rude jests about their generals in triumphal processions, to avert the evil eye from them in their great glory.

5-6. Take the second line first. 'I ask you to read (jussive subjunctive) my poems with the same smile (lit. forehead) with which you look at....' Thymele was a dancer, Latinus an actor.

7. **censura.** Domitian made himself perpetual censor, which included censorship of literature, in 85.

8. 'My books are sportive, but my life is pure.'

nobis = *mihi*.

IV. (I, 13)

1. **suo Paeto**] 'to her beloved Paetus'.

2. **de visceribus suis**] 'from her own heart'.

3. **Si qua fides**] Supply *est mihi*: 'if you believe me (lit. if there is any belief), the wound which I have inflicted (on myself) does not hurt, but the one which you are going to inflict (on yourself) does hurt me'. When Caecina Paetus was involved in the conspiracy of Scribonianus, governor of Dalmatia, against Claudius in A.D. 42, his wife, the elder Arria, stabbed herself first and then handed the dagger to her husband with the words: '*Paete, non dolet*', 'It doesn't hurt, Paetus'. For this and other instances of her devotion and courage see Pliny, II.

V. (1, 70)

1. **salutatum**] Supine of purpose after a verb of motion: 'to present my greetings' as a *salutator* or early morning caller at the house of a great man.

2. **nitidos...lares**] 'the splendid house', lit. the glistening household-gods.

officiose] 'to pay your respects'=*officii causa*. It is 'attracted' into the vocative case.

3. **Quaeris iter**] Supply *si*: 'if you ask the way I will tell you'. Martial lived on the Quirinal, Proculus on the Palatine.

Castora is Greek accusative singular, 'you will pass by (the temple of) Castor which is next to (the temple of) hoary Vesta'. The name of a deity is often used for its temple. 'The virgins' home' was the abode of the Vestal Virgins. All these buildings were just south-east of the *Forum Romanum*.

5. 'Then you will make for the hallowed Palatine (Hill) along the Sacred Slope, where many a statue of the supreme emperor (*ducis*) gleams.' The *Clivus Sacer* was part of the *Sacra Via*: the Palatine was 'hallowed' because it contained several temples and in particular the huge imperial palace of Domitian. The final *-o* of *imago* is short.

8. **moles**, which is inside the relative clause, is the subject of the main verb. 'Let not the rayed mass of the marvellous Colossus delay you, which is proud to surpass the statue (*opus*) at Rhodes.' Nero placed a huge statue of himself in his Golden House, but Vespasian substituted for Nero's head one surrounded by the rays of the sun, like that of the Colossus of Rhodes, and moved it to the Sacred Way. Hadrian later moved it to the entrance of the amphitheatre, which was hence called the Colosseum. The statue was 119 feet high, the Colossus at Rhodes 105.

9. **Flecte vias hac qua**] 'turn your steps (lit. ways) by the way where...'. There was a temple (*tecta*) of 'wine-soaked Bacchus' on the Palatine.

10. **Cybeles** is genitive: 'and where the dome of Cybele stands with painted (figures of) Corybants'. This was the temple of Cybele on the Palatine, presumably decorated internally or externally with figures of Corybants.

11. **Protinus...**] 'immediately after this, on the left, there is a house with a gleaming façade (lit. household gods bright in front), and the hall...which you must approach'.

13. **nec metuas fastus**] A prohibition: 'and do not fear its pride'.

14. Nulla magis...] 'no door stands more widely (*magis*) open, with both its leaves (flung back)', lit. with its whole door-post. The doors were double, with a door-post on each side.

15. Nec propior (est)...] 'nor is there one which Phoebus loves more dearly', lit. a nearer one that he would love. *amet* is 'generic' or consecutive subjunctive, 'the kind of house that...'. The 'sisters' are the Muses.

16. Quare] 'why?' **Ipse]** 'your master', i.e. Martial.

17. Sic licet excuses] 'you may give this excuse' = *licet tibi sic excusare.*

Quia...] 'because, whatever these verses are that you read (lit. whatever are read), a morning caller could not have written them', i.e. a poet would not have the time to waste on paying morning calls.

VI. (I, 88)

1. raptum domino...] 'snatched from your master (dative of indirect object) in your youthful years'.

2. The *via Lābīcāna* ran S.E. from the Esquiline gate and, like most main roads, was lined with tombs for several miles.

3. Accipe non...] 'receive from me, not the tottering weight of Parian stone...', lit. weights tottering with Parian stone. Parian marble was famous.

4. Quae cineri...] '(a monument) soon destined to fall (*ruitura*) which wasted labour gives'.

5. faciles] 'pliant'.

6. quaeque (= *et quae*): 'and grass that grows green, bedewed with my tears'.

8. perpetuo tempore] 'for ever'. Note this Silver Latin use of the ablative instead of the accusative for extent of time.

9. Cum mihi...] 'when Lachesis (one of the Fates) has spun to the end (lit. shall have spun) my last years, not otherwise do I give orders that my ashes should lie'. Note the short final *-o* of *mando*, and the infinitive instead of *ut* with the subjunctive.

VII. (I, 109)

1. passere nequior Catulli] 'more mischievous than the sparrow of Catullus'. Catullus wrote a poem about a pet sparrow belonging to his sweetheart Lesbia.

4. carior] 'more precious.'

5. 'Issa is a little dog, the pet of Publius'.

6. queritur] 'whines'.

8. Collo nixa...] 'she lies resting against her master's neck, and sleeps (so lightly) that her breathing is not heard'.

17. Hanc ne lux...] 'so that her last day may not take her away altogether (*totam*), Publius is painting her (or, is having her painted) in a picture'.

19. In qua...] 'in which you will see Issa so lifelike that not even she herself is so like herself'. *nec* here = 'not even'.

21. denique] 'in fact'. **pone]** 'compare'.

22. veram, 'alive' or 'real'.

VIII. (I, 117)

1. nobis = *mihi.*

2. Vis mittam] 'may I send a slave?' lit. are you willing that I should send: jussive subjunctive.

3. Cui tradas is a final relative clause: 'to whom you may hand...'.

epigrammaton is Greek genitive plural.

4. Lectum quem...] 'which I shall send back to you as soon as I have read it'.

5. Non est quod...] 'there is no reason why you should...'; *quod* is adverbial accusative and the verb is indirect deliberative subjunctive. See Pliny I, 5.

6. Longum est (iter)] 'it's a long way, if he should wish to come to the Pear Tree'. Note the mixture of moods in the conditional sentence. *Pirus* was a district of Rome, probably named after a pear-tree that stood there before the city spread.

7. scalis...altis] 'up three flights of stairs, and high ones too'. Martial lived in an *insula,* or block of flats; see p. 15.

8. 'You can obtain what you want nearer home.'

petas licebit = *licebit tibi petere.*

9. Argi Letum] See note on II, 1. This was the district where booksellers had their shops. Martial divides this word (by 'tmesis') to show one derivation, 'the death of Argus'; others derive it from *argilla,* 'clay'.

soles subire] 'you often go down to...'.

10. Contra governs *forum*: 'opposite Caesar's forum there is a shop with its door-posts on each side (*hinc et inde*) all covered over with book titles', lit. all written over; this was a usual kind of advertisement at booksellers' shops.

12. perlegas] 'read through the names of...'. The *-o* of *cito* is short.

13. nec roges...] A command: 'and you needn't ask Atrectus (for a copy of my book, because as soon as he sees you) he will give you a Martial for five denarii, from the first or the second shelf', lit. nest or pigeon-hole, in which the rolls were kept; the papyrus rolls were often 'smoothed with pumice-stone and adorned with purple'. A *denarius* was a silver coin worth about 10*d.* but with a much higher purchasing power in those days; *Denaris* is contracted ablative (of price). Five denarii was the price of an expensively-finished presentation copy; ordinary books were often much cheaper.

18. Tanti non es?] '"you are not worth it", you say?' *tanti* is genitive of value.

Sapis] 'you are a shrewd fellow'. Martial means that a man so mean as Lupercus ought not to waste his money on buying books when he can borrow them.

IX. (II, 75)

1. Verbera...] 'a lion, accustomed to bear the blows of his fearless keeper, and to endure tamely a hand thrust into his mouth, suddenly forgot his peaceful ways (*pacem*) when his fierceness returned, such as ought not to have been shown (*esse*) even on...'. The *-o* of *leo* is short. *nec* again = 'not even'.

5. duo...puerilia corpora ('two boys') is governed by *peremit* in l. 7.

de tenera...turba] 'from the youthful band'.

6. renovabat] 'was smoothing over', to remove the blood-stains from the sand during the games in the amphitheatre.

7. Saevus et infelix] 'the savage and ill-starred creature', both because of the act and its consequences.

8. Martia harena] 'the sand of the Roman arena'; see II, 4.

9. Exclamare libet] 'one may exclaim', lit. it pleases to exclaim.

10. 'Learn how to spare children from our own Roman she-wolf', i.e. the one that nursed Romulus and Remus.

X. (II, 87)

1. amore tui (objective genitive) **ardere]** 'burn with love for you'.

2. Qui] 'you who...'. **natantis]** 'of a man swimming...'.

XI. (III, 5)

1. Vis commendari...] 'now that you are so eager to hasten (*cursurus*) to the city without me, do you wish to be introduced to many people, or will one be enough?' Martial was absent in

Cisalpine Gaul at Forum Cornelii, near Ariminum in North Italy, when this book was published.

3. **cui non eris hospes]** 'to whom you will not be a stranger'.

4. 'Julius, a name constantly on my lips'. *nomen* is in apposition to *Iulius*.

5. **primae...Tectae]** 'on the very entrance of the Covered Way', lit. on the threshold of the first (part of the) Covered Way. Supply *Viae* with *Tectae*. There were two such colonnades, one in the north of Rome near the *Mausoleum Augusti*, the other in the south running from the *Porta Capena* to the temple of Mars. It is not certain which one is meant here.

6. 'He lives in the house where Daphnis once lived.' This Daphnis is unknown.

7. **Est illi coniunx]** 'he has a wife'.

manibusque sinuque] 'with hand and heart'.

8. **vel...eas]** 'even if you were to arrive all dusty'. The *si* is omitted.

9. **Hos tu seu...]** 'whether you see them together, or her or him first, you will say, "Marcus (i.e. Martial) sends you greeting"', lit. bids (you) hail. *videris* is future perfect.

11. **alios...]** 'let others be introduced by a letter; he who thinks that he need be introduced to his friends is wrong'.

XII. (III, 19)

1. 'A bear is on show next to the Hundred Columns', which may have been part of Pompey's Theatre, in the *Campus Martius*.

2. **fictae qua...ferae]** 'where images of wild-beasts...'. *platanona* is Greek accusative singular.

3. **patulos...hiatus]** 'was exploring (or 'challenging') in play its gaping jaws'.

4. **mersit in ora]** 'thrust down its throat'. The name Hylas is taken from that of the beautiful youth who was drowned while accompanying Hercules on the Argonautic expedition.

5. **caeco in aere]** 'in the dark depths of the bronze'.

6. 'And was living there with a life more deadly than (that of) the bear itself', lit. worse than the beast, an instance of 'brachylogy' or shortened form of comparison.

7. **Non sensit...]** 'did not realize that the hidden danger was there, except when he had received the bite (lit. tooth) and was dying'.

8. **o facinus]** 'oh what a pity (lit. crime, accusative of exclamation) that the bear was a false one', because the bear would not have had the chance to kill the boy if it had been alive.

XIII. (III, 38)

1. 'What reason or what confidence brings you to Rome?'

2. **inde**] 'from there'. **refer**] 'tell me'.

3. **Causas agam...**] 'I will plead causes (in court) more eloquently than Cicero himself'.

4. **in triplici foro**] 'in the three forums', i.e. *Forum Romanum, Forum Iulium, Forum Augusti.*

6. **Noras** = *noveras*: 'you knew'. The -*o* of *pensio* is short.

neutri...] 'neither earned his full rent', by his eloquence.

7. **hinc**] 'from this profession'.

pangentur carmina nobis] 'I will write poems'.

nobis (= *mihi*) is dative of the agent, generally used after a past participle passive or a gerundive.

8. **Audieris** is future perfect: supply *cum*: 'when you have heard them, you will say that they are Vergil's work'.

9. **Insanis** is a verb.

omnes...] 'in all those who stand there in their threadbare (lit. chilly) cloaks you see Ovids and Vergils', i.e. all poets think that they are great writers.

11. **Atria magna colam**] 'I will court the halls of the great', and live as a client.

ista res] 'that method'.

13. **Quid faciam?**] Deliberative question: 'What am I to do? advise me: for I am determined (*certum est*, supply *mihi*) to live at Rome'.

14. **casu vivere potes**] 'you may perhaps pick up a chance living', lit. live by chance, meaning that only a bad man is sure of prosperity. *bonus* here means 'honourable'.

XIV. (IV, 8)

1. 'The first and second hours of the day weary the morning-callers', who had to come early to the levée of their patron and perhaps wait a long time, dressed in the heavy toga, until he was pleased to receive them. The Roman day from dawn to sunset was divided into twelve hours, which varied in length with the seasons.

2. **exercet raucos**] 'keeps advocates busy till they are hoarse'. *raucos* is a 'proleptic' or anticipatory adjective; they are not hoarse until they have spoken for a long time.

3. **In quintam**] 'until the fifth hour'.

4. Sexta (est) quies] 'the sixth brings rest to the weary', i.e. it was the midday siesta.

finis] 'the end of business hours'.

5. 'The eighth until the ninth is the time (lit. suffices) for the oily wrestling-schools', where men oiled their bodies after exercise; the bath followed, but is not mentioned.

6. Imperat here takes the construction of *iubet*: 'bids (us) disturb the piled-up cushions', at the dinner table, i.e. by lying on them.

7. Martial asks Euphemus to recommend his book to Domitian after dinner was over, when the wine was circulating.

8. tua cura] Because Euphemus had general care of the meal and wine.

9. nectare, and **ambrosias** in l. 8, suggest flatteringly the divinity of Domitian. *laxatur* means 'unbends'.

10. Ingenti manu] 'in his all-powerful hand'.

parca] Domitian was temperate in drink.

11. gressu timet...] 'my Muse (i.e. book) fears to go with sportive step to Jupiter (i.e. the emperor) in the morning', when he was not mellowed by wine.

XV. (IV, 14)

1. 'O Silius, glory of the Castalian sisters', i.e. the Muses, to whom the Castalian spring on mount Parnassus in Greece was sacred.

2. qui periuria...] 'you who are suppressing (i.e. describing as being suppressed) with mighty utterance (*ore*) the perjuries of barbarian rage'. This refers to Silius' poem on the second Punic War, which still survives. 'Punic perfidy' was proverbial at Rome.

4. Astus is accusative plural, governed by *cogis*.

leves] 'false'.

5. Africanis] Scipio Africanus Major, the conqueror of Hannibal at Zama in 202 B.C., and his adopted grandson, also surnamed Africanus, who destroyed Carthage in 146, though Statius' poem ends with Hannibal's defeat and does not include the exploits of the younger Scipio. The plural may mean 'men like the great Africanus', without the anachronism.

6. 'Lay aside your sternness a little'; translate the ablative absolute as an imperative.

7. Dum blanda...] 'while December, sporting idly (*vagus*) with the seductive hazard, echoes on all sides with the uncertain dice-boxes'. Dicing was permitted legally only during the festival of the Saturnalia, 17–21 December.

9. tropa is an adverb: 'and plays "tropa" with the more treacherous knuckle-bones'. In the game 'tropa' the *talus* was thrown by hand and could give scope for skill or cheating, while it was impossible to control play from the dice-box.

10. commoda] Imperative: 'give your leisure to my muses', i.e. read my verses during the holiday.

11. Nec negatives *torva* alone: 'and read my books, which are steeped in sportive jests, with a brow not stern but cheerful', lit. relaxed. *madidos* also gives a suggestion of drunkenness.

13. tener] 'the love-poet'. Catullus wrote a poem on Lesbia's sparrow (see VII, 1), but though Martial compares himself to Catullus and Silius to Vergil, it was an anachronism to connect Catullus with Vergil, who was only sixteen when Catullus died. *passerem* here means 'his poem "The Sparrow"'.

XVI. (IV, 30)

1. Baiano a lacu] 'the lake at Baiae' was a *piscina* or fish-pond belonging to Domitian.

2. ne nocens recedas] 'so that you may not have to depart with guilt on your head', for having desecrated the pool by fishing there.

3. natantur] A passive use: 'are swum by sacred fish'.

4. norunt = *noverunt*: 'know', a perfect form with present meaning. *nosco*, the present tense, means 'get to know'.

5. qua nihil...] 'mightier than anything else on earth', lit. than which nothing is greater.

6. Quid, quod...] lit. what of the fact that...? tr. 'moreover, they even have their own names, and each one comes to his master's voice when summoned'. *ad* of course governs *vocem*.

8. Hoc profundo] Local ablative without *in*: 'at this deep pool'. *Libys* is 'a Libyan fisherman'.

9. tremente (= *trementi*): 'with quivering rod'.

10. 'Suddenly (struck) blind with the loss of his eyes', lit. his eyes being snatched away.

13. Baianos...rogator] 'sits as a beggar beside the Baian pools'.

14. innocens] 'while still guiltless'.

15. simplicibus] 'harmless', i.e. not attached to a hook.

16. venerare is imperative: 'respect these dainty fish'.

XVII. (IV, 64)

2. **Hortis...beatiora**] 'more blessed than the gardens of the Hesperides'. These were situated on an island beyond Mount Atlas and contained golden apples; Perseus visited them, and Hercules carried off the apples as one of his labours.

3. **Longo iugo**] Local ablative without *in*.

4. 'A broad retreat overlooks the (surrounding) hills.' The villa was built on a level space on a ridge of the Janiculum (a mile west of Rome) higher than the neighbouring hills.

5. **planus modico...**] 'the flat summit with its gentle rise enjoys a purer air (*caelo*)'.

7. **nebula tegente** is ablative absolute: 'and when the mist covers the hollow valleys it (i.e. the summit) alone shines in a light that is all its own'.

9. **Puris leniter...**] 'the graceful roof of the lofty villa rises gently towards the unclouded stars'.

11. **Hinc**] 'from here', *illinc* (l. 18) 'from here again'.

dominos videre...] 'one may (*licet*) see the seven sovereign hills (see II, 3) and take in at a glance (*aestimare*) the whole of Rome'.

14. **Et quodcunque...**] 'and all the cool retreats that lie near the city', lit. whatever coolness lies.

15. **breves**] 'tiny'. *Rubrae* is the town of Saxa Rubra.

16. **virgineo cruore**] Anna Perenna was worshipped on the Ides of March, but what is the reference to a virgin's blood is unknown; human sacrifices to Anna Perenna are nowhere mentioned.

18. **Flaminiae...**] 'the traveller on (lit. of) the Flaminian and the Salarian Way can be seen (*patet*), though his carriage makes no noise (ablative absolute), so that the (sound of his) wheels may not be burdensome to gentle sleep, which neither...'.

23. **Cum**] 'although'.

Mulvius] Supply *pons*. The Mulvian Bridge carried the Flaminian Way across the Tiber about two miles north of Rome.

24. 'And the ships glide swiftly down the sacred Tiber', lit. fly gliding.

25. **hoc rus...**] 'this country estate—or rather it should be called a town house—its owner places at your disposal: you will imagine it to be your own'.

27. **Tam non invida (est)...**] 'so ungrudging is it...with such friendly hospitality does it lie open to you'.

29. Credas] Potential subjunctive: 'you would believe it to be the kindly home of Alcinous (the king of Phaeacia, who entertained Ulysses) or of Molorchus recently (*modo*) enriched'. Molorchus was a shepherd who entertained Hercules on his quest for the Nemean lion and in return was given land by him.

31. omnia parva qui putatis] 'you who think all estates small', because they are never satisfied.

32. centeno ligone] 'with a hundred hoes', i.e. workmen: collective singular. *centeni* often = *centum*, as in XII, 1.

domate] 'cultivate'.

34. dedite...] 'give up Setia on its hill to a single husbandman', i.e. make it all one estate. Setine wine was famous.

35. Dum me iudice...] 'provided that in my opinion (ablative absolute) the few acres...are preferred to those estates of yours'.

XVIII. (V, 34)

1. From this epigram it is thought that Fronto and Flaccilla were Martial's parents. 'I hand over to your care, Fronto my father and Flaccilla my mother, this maiden, who is my pet (lit. kisses) and darling'. *oscula* and *delicias* are in apposition to *hanc puellam*. The final *-o* of *Fronto* is short.

3. ne...horrescat] Because she would be under their protection in Hades. Fronto and Flaccilla were dead.

4. Tartarei canis] 'the hound of hell' is Cerberus, who had three heads.

5. Inpletura fuit...] 'she would just (*modo*) have completed her sixth cold winter (lit. the cold of a sixth winter) if she had not lived as many days too few (*minus*)'. She was within six days of her sixth birthday.

7. Inter tam veteres...] 'by the side of (*inter*) such aged protectors may she sport and play'.

9. non rigidus (= *levis*) go together: 'may the turf lightly cover her tender bones, and do not lie heavy upon her, O earth; she was not heavy to you'. *nec fueris* is a negative command.

XIX. (VII, 19)

1. Fragmentum quod...] 'the fragment which you think to be...this was (part of) the first ship to traverse the unknown sea'. *haec* is 'attracted' from the neuter to agree with *carina*.

3. Cyaneae ruinae] 'the deadly clashing together of the

Cyanean Rocks'. These were the Symplegades, at the mouth of the Bosphorus, said to clash together and crush vessels.

potuere = *potuerunt*. The antecedent of *quam* is *eam* understood, which is the object of *vicerunt* in l. 5. 'Time (lit. ages) has overcome the ship which neither...'.

4. **tristior**] 'more baneful'. The Scythian Sea is the Black Sea. The Golden Fleece was at Colchis on the shores of the Black Sea, and was the object of the Argonautic expedition.

5. **cesserit annis**] 'it has yielded to the passage of the years'.

6. 'This little plank is more sacred than the vessel itself was when it was whole', because of its antiquarian value.

XX. (VII, 61)

1. **Abstulerat...**] 'had robbed us of the whole city'.

2. 'No threshold remained within its own bounds', but gradually encroached on to the street.

3. **Germanice** is Domitian, who took the title from his victory over the Chatti in A.D. 83.

tenues crescere vicos] 'you ordered the narrow streets to increase in size'.

4. **modo quae fuerat**] 'what was lately a footpath has become a highway'.

5. To attract trade, the doorposts of wine-shops were hung with flagons, chained together to prevent their theft.

6. **medio luto**] Local ablative without *in*: 'to walk in the mud in the middle of the street'.

7. Take *nec* first: 'nor is a razor drawn wildly (lit. blind)...'. Barbers used to shave customers in the middle of the street, brandishing their razors to the public danger.

8. **nigra**] 'grimy'. *nec...aut* = *nec...nec*.

9. **servant**] 'now keep within...'. The final *-o* of *copo* is short.

10. **nunc Roma est**] 'the city is now Rome, lately it was...'.

XXI. (VII, 96)

1. 'Mourned by Bassus (lit. the grief of Bassus), I lie buried here, the child Urbicus...'.

2. **nomen**] Because Urbicus is derived from *urbs, genus* because he was a Roman.

3. **de prima...trieteride**] 'were lacking to complete (lit. from) my first three years of life'. *derant* (= *deerant*) is from *desum*.

4. 'When the stern goddesses cruelly broke my mortal thread'. The three Fates were said to spin the thread of a person's life and break it off at his death. *pensum* really means 'wool "weighed out" for a slave to spin'.

5. **Quid...mihi...profuit?**] 'what good did my...do to me?'

lingua] 'my childish prattle'.

6. **Da lacrimas**] 'shed tears over my tomb, you who read this inscription'.

7. **Sic...**] 'so may he, whom you wish to survive you (i.e. your heir), not go to the waters of Lēthē (in the underworld) except when older than Nestor', lit. later than Nestor, the king of Pylos, who in Homer lived for three generations.

XXII. (VIII, 69)

1. **veteres solos**] 'only ancient poets'.

2. **Nec...nisi mortuos**] 'and...none but dead ones'.

3. **Ignoscas petimus**] 'I ask you to excuse me'. *Ignoscas* is jussive subjunctive. *petimus = peto.*

tanti] Genitive of value: 'it is not worth while to die, merely to please you'.

XXIII. (IX, 61)

1. **domus est notissima**] 'there is a well-known house...'.

2. **Qua...**] 'where Corduba shows its love for (lit. loves) the Baetis', by being built on the river.

Baetin is a Greek accusative; it is the modern Guadalquivir.

3. 'Where the fleeces are pale yellow (*flava pallent*) with native ore, and living gold covers (lit. smears) the western flocks'. Martial compares the natural colour of the Spanish wool to the gold of the Golden Fleece.

5. **mediis**] The inner court (*peristylium*) of a house was open to the sky and often had trees planted in it.

totos amplexa penates] 'overshadowing the whole building', lit. embracing the household gods.

7. 'Which the auspicious hand of that unconquerable guest planted.' Caesar was in fact never defeated in a pitched battle.

8. 'The twig began to grow from (the touch of) that hand.'

9. **Auctorem...sentire videtur**] 'it seems to recognize its creator and lord'.

10. **Sic viret**] 'so green it is'. 11. **madidi**] 'tipsy'.

12. **sera**] 'late at night'.

13. **nocturnum Pana** (Greek accusative singular): 'night-wandering Pan'.

14. Scan carefully. *rustica* is nominative.

15. 'The household gods have often been steeped in fragrance (=*oluerunt*) when Bacchus held revel': ablative absolute.

16. **Crevit et...**] 'and the foliage (lit. shade) grew more luxuriantly from the wine poured out (upon its roots, possibly as a libation)'.

17. **rubens deiecta est herba**] 'the grass has been scattered and reddened with yesterday's garlands'. *rubens* is here 'proleptic' (see XIV, 2).

18. **suas...dicere...rosas**] 'say which roses were his'. Guests used to pelt each other with roses. The -*o* of *nemo* is short.

19. **dilecta deis**] 'beloved by the gods'. *deis* is dative, 'dear to the gods', or it may be dative of the agent.

20. **Ne metuas**] A command: 'do not fear'. It will never be cut down by the axe (*ferrum*) and burnt. Julius Caesar was *divus*, 'deified', so flames (*focos*, lit. hearths) that burnt this tree would be sacrilegious.

21. **Perpetuos...**] 'you may hope for everlasting honour for your leaves; it was not the hands of Pompey that planted you'.

posuere=*posuerunt*. Pompey was defeated by Caesar at Pharsalus in 48 B.C., but the conqueror's tree will flourish for ever.

XXIV. (IX, 68)

1. **Quid tibi nobiscum est?**] 'What have you to do with us?' A *ludus* was an elementary school where the three R.s were taught; a *grammaticus* then taught languages and literature, and a *rhetor* finished the education with rhetoric.

2. **Invisum... caput**] 'creature loathed by...'.

3. **rupere**=*ruperunt*. *silentia*, plural, often=*silentium*.

4. 'You make a thunderous noise with your savage voice and with the strokes of your cane.' Discipline was strict in Roman schools.

5. **Tam grave** (adverbial accusative)...] 'so loudly does the bronze resound on the smitten anvils when a smith fits (the statue of) a lawyer astride his horse', lit. to the middle of his horse. Successful lawyers often had equestrian statues of themselves made.

7. **Mitior**] 'less deafeningly'.

8. 'When his own supporters urge on the winning gladiator', lit. shield. This was the 'Thracian' gladiator, armed with a light buckler and scimitar against a more heavily armed opponent. *sua* here refers to the object, not to the subject, as is usual.

9. **Vicini**] 'we neighbours'. *tota nocte* is used where strictly the accusative is required to express duration of time, but Silver Latin authors often use the ablative, as in VI, 8.

10. 'It is a small thing to lie awake (occasionally), but a serious matter to be awake all night.' An intentional exaggeration, unless Martial means that he went to bed so late that the school had already started in the early morning.

11. **Vis (tantum)...quantum...**] 'are you willing to receive as much to be silent as you do to shout?'

XXV. (IX, 70)

1. **O mores! o tempora!**] 'oh the manners! oh the times!' exclamatory accusative. *Tullius* is Cicero, who said this in his speech against the revolutionary Catiline in 63 B.C.

3. **Cum gener...**] '(and again) when son-in-law and father-in-law...'. Julius Caesar's daughter Julia married Pompey, and after her death the Civil War broke out. Cicero is not known to have made this remark again.

6. **Quod tibi...**] 'what is there that does not please you?' *quod placeat* is a 'generic' subjunctive.

7. **Nulla ducum...**] 'there is no fierceness of leaders, no madness of the sword'.

8. **licet**] Supply *nobis*: 'we may enjoy...'.

9. **Non nostri (mores)...**] 'it is not *our* "manners" that cause (lit. make it that...) your "times" to seem disgraceful, but *your own* "manners" cause it'.

XXVI. (IX, 101)

1. 'O Appian Way, greatest glory of Italian roads (*viae*, the singular means "road-making") which Caesar hallows, to be worshipped in the form of Hercules', lit. in a similar Hercules. Domitian built a temple for Hercules, in which he placed a statue of the god with the features of the emperor.

3. **Alcidae prioris**] 'of the former Hercules', as opposed to his re-incarnation, Domitian.

4. **Disce**] 'listen to me'. *Libyn*, Greek accusative, is the

giant Antaeus; the golden apples came from the garden of the Hesperides.

5. 'He stripped of her Scythian girdle the shield-guarded Amazon': Greek accusative. This was the Amazon queen, Hippolyte. *nodo* is ablative of separation.

6. 'He added (the capture of) the lion's skin (lit. backs; this was the Nemean lion) to (the slaughter of) the Arcadian boar.'

7. **silvis** and **astris** ('from the sky') are dative of indirect object. *Stymphalidas*, supply *aves*; Hercules roused them with a rattle and then shot them down with his bow and arrows.

8. **a Stygia...**] 'he returned from the waters of Styx with the dog Cerberus'.

9. **vetuit...**] 'he prevented the self-productive hydra from getting fresh strength from being slain', lit. from deaths. The hydra was a many-headed water snake, which grew two more heads for every one cut off.

10. **Hesperias boves**] i.e. the oxen of Geryones. The 'Tuscan river' is the Tiber, over which he drove the oxen on his way back from Spain to Greece.

In this list ten labours are mentioned, but of the traditional twelve three are omitted—the cleansing of the Augean stables, the capture of the Cretan bull and of the mares of Diomedes, and the killing of Antaeus is given instead.

11. **Haec minor...(gessit)...**] 'the lesser Hercules did these exploits: hear what the greater one (i.e. Domitian) did'.

12. **Sextus ab Albana...**] 'whom the sixth mile-stone from the Alban citadel celebrates'. It was at this point on the Appian Way that Domitian built the temple of Hercules, as mentioned in l. 1. The 'Alban citadel' was Domitian's country-house at Alba Longa, about fourteen miles from Rome.

13. **Adseruit...**] 'he freed the Palatine when it was seized by an evil power'. In A.D. 69 (the year of the Four Emperors) Vespasian was absent from Rome, and young Domitian helped to capture the Palatine from the Vitellians and secure the throne for his father.

14. **suo pro Iove puer**] 'when a boy, in defence of his patron deity Jupiter'. During the same fighting in 69 Domitian was besieged in the temple of Jupiter on the Capitol. The god is called 'his patron' because of this, and because Domitian restored the temple after a fire which took place in the reign of Titus.

15. **Solus**] He was the only member of his family in Rome at the time.

Iuleas habenas] 'the reins of Julian power', i.e. of Julius Caesar, though Domitian was not of the Julio-Claudian line.

cum] 'although'.

16. **Tradidit**] 'he handed over (the power to his father)'.

tertius] After his father Vespasian and his brother Titus.

In suo orbe] 'in the world which was already his own', because he had held the city *solus* (l. 15).

17. 'To break the horns of a river' was to subdue the people living on its banks, as river-gods were often represented as having horns. Domitian defeated the Getae (Dacians) and Sarmatae, north of the Danube, in 89 and 92. *ter* is an exaggeration.

18. **lavit**] 'plunged'.

19. 'Unwilling to lead (*duxisse* is poetic for *ducere*) triumphs which he often refused'; another flattering exaggeration.

20. **Victor**] 'as conqueror'. He took (*tulit*) the name Germanicus to celebrate a victory over the Germans in 83.

21. **mores populo**] 'morals to his people'. He was *censor morum.*

otia ferro] 'peace to the sword'.

22. **Astra suis**] 'immortality (lit. stars) to his own family', by consecrating temples to the Flavian *gens* and making his family divine. *caelo sidera* seems to mean the same thing: 'fresh stars to the sky'. *serta Iovi,* by dedicating his Sarmatic laurels to him.

23. 'The divinity of Hercules is not enough for such deeds: let this god lend his features to the Tarpeian Father', i.e. let Domitian be worshipped in the form of Jupiter on the Capitol (the Tarpeian Rock).

XXVII. (X, 30)

1. **Formiae** is vocative plural and *dulce litus* in apposition, so translate: 'O temperate Formiae, pleasant shore'. It was on the coast of Latium, about 70 miles south-east of Rome.

2. **oppidum Martis** is Rome, as in II, 4. The subject of *fugit* is *Apollinaris*, and *vos* is governed by *praefert.*

5. **Non ille...miratur**] 'he does not admire so much his chaste wife's sweet Tibur, nor the quiet retreats of Tusculum or Algidum, nor Praeneste and Antium'. These places were all beauty spots in central Italy; possibly Apollinaris' wife was born at Tibur. Take *sic* with *miratur*, and *nec* (l. 7) before *Praeneste.*

8. Circe the witch was supposed to live at Circeii; Caieta was named after the Trojan nurse of Aeneas.

9. Desiderantur (ab eo)] tr. 'he does not long for seductive Circe nor...'. Marica was a nymph who had a sacred grove on the river Liris at Minturnae.

10. 'Nor Salmacis, bathing in the Lucrine stream', probably the name of a spring which ran into the Lucrine lake.

11. Hic] 'here at Formiae'. Thetis was a sea-nymph, so *summa Thetis* means 'the surface of the sea'.

12. viva quies] 'the living slumber', i.e. the almost unbroken calm.

13. adiuvante aura] Ablative absolute: 'with the aid of the breeze'. *phaselon* is a Greek accusative.

14. 'As refreshing coolness comes when the purple fan of a maiden who does not like the summer heat is moved'.

16. longo in mari] 'far out to sea'.

18. Spectatus alte...] 'a fish seen from above takes the line which is cast from...'. The villa must have been built partly on moles projecting into the sea.

19. Si quando] 'if ever'. Nereus was a sea-god, Aeolus the god of the wind. The *-o* of *quando* is short.

20. tuta de suo mensa] 'the table, safely supplied from its own resources'. The fish-ponds produce fish for the table independently of the sea.

21. lupos vernas] 'home-bred pike'.

23. Nomenculator] Usually a slave who reminded his master of people's names, but here the keeper who knew the name of every fish. See also xvi, 6.

24. 'Aged mullets come forward when told to appear.'

25. 'When do you allow (their owner) to enjoy such pleasures?' *Roma* is vocative.

26. Quot Formianos...] 'how many days at Formiae does the year grant to (lit. put down to the account of) him when he is tied (*haerenti*) to the business of the city?'

29. Dominis parantur...] 'those joys are provided for your masters, but they belong to you', lit. serve you.

XXVIII. (X, 47)

2. haec sunt...] 'these are the things which make...'.

3. Res non parta (from *pario*)...] 'wealth, not earned but inherited'.

4. Non ingratus ager...] 'land which yields a good return, an ever-blazing hearth'.

5. **Lis numquam**] 'no lawsuits'; *numquam* is used almost like an adjective; perhaps 'never a lawsuit' would do.

toga rara] 'the toga seldom worn'. The dislike of the heavy formal toga was universal.

6. **Vires ingenuae**] 'a gentleman's constitution', not that of a mere labouring slave.

7. **Prudens simplicitas**] 'tactful frankness'.

pares] 'well-matched'.

8. **Convictus facilis**] 'congenial company', lit. easy social intercourse.

sine arte mensa] 'a plain table', without fancy dishes.

9. **Nox non ebria**] 'evenings not spent in drunkenness'.

11. **qui faciat**] 'Generic' subjunctive: 'such as makes the hours of darkness short'.

12. **Quod sis...**] 'the wish to be what you are and to prefer nothing else'. *velis* = *ut velis* or *velle* and is equivalent to a noun. *sis* is 'generic' subjunctive: 'the kind of person you are'.

13. **nec metuas** is another noun like *ut velis*: 'neither the dread of your last day nor the wish for it'.

XXIX. (X, 103)

1. **Augusta...creat**] tr. 'born at Augustan Bilbilis on its bracing (*acri*) hill-side', lit. whom Bilbilis produces. It is perhaps called *Augusta* as being a Roman colony. *mihi* is dative of advantage, which is best translated here as though it went with *municipes*, 'my fellow-townsmen'. The *-o* of *Salo is* short.

3. **Ecquid** is adverbial accusative: 'does the glory...delight you at all?'

4. **sumus** = *ego sum*: 'I am your honour...'.

5. **sua** refers to *Catullo*, not to the subject of the sentence: 'his own Verona does not owe more to the elegant (or perhaps 'the love poet') Catullus (than Bilbilis owes to me), and she (*illa*, i.e. Verona) would like me to be called her (son) no less (than him)'. Verona was Catullus' birthplace. *velit* is 'potential' subjunctive.

7. 'A thirtieth summer has been added to four harvests since (*ut*) you have been offering (lit. are offering)...', i.e. it is 34 years since Martial left Bilbilis.

9. **dum colimus**] 'while I dwelt in...'.

dominae] 'imperial'.

10. **Mutavere** (= *mutaverunt*): 'the realms of Italy turned my hair white', lit. changed it.

11. 'If you receive (me) on my return (*reducem*, from *redux*) with kindly hearts, I come to you.' The present tenses are used instead of the future.

12. si geritis] 'if you have unfriendly hearts'.

licet] Supply *mihi*: 'I can go back (to Rome)'. He is afraid that his countrymen may be jealous of him.

XXX. (X, 104)

1. nostro comes Flavo] 'with my friend Flavus', lit. as companion to him.

2. faventis undae is genitive of quality depending on *mare*: 'over the distant sea, but one with friendly waters'.

3. tuis ventis] 'with favouring winds', lit. your own.

4. pete arces] 'make for the heights'.

5. te rota tollet...] 'a carriage (lit. wheel) will take you, and carried speedily along you will see...perhaps in the fifth stage', lit. carriage, i.e. after the fourth change of horses.

8. Quid mandem...] 'do you ask what instructions I give you? That you should greet (*salutes*)...'.

9. ante...visos] '(last) seen by me (dative of agent) thirty-four winters ago'. *ante brumas* = *brumis ante*.

11. Ipsa protinus a via] 'straight from the road itself', i.e. as soon as you leave the carriage, or perhaps, without waiting to leave the carriage.

13. nec laboriosos] 'and not difficult to keep up'.

14. Secessus pretio paret salubri] 'to buy me a...country-retreat at a moderate (*salubri*) price'. *paret* is jussive subjunctive depending on *admoneas*.

15. Qui faciant] 'Generic' subjunctive: 'such as will make your author lazy'.

16. Haec sunt] Supply *ea quae mando*: 'these are my instructions'.

tumidus magister] 'the hot-tempered captain'.

17. aura portum laxavit melior] 'a fair (lit. better) breeze has opened the harbour (for your departure)'.

19. 'One (passenger) does not keep a ship waiting, I suppose you know', so hurry on board or you will be left behind. The *-o* of *puto* is short.

XXXI. (XI, 18)

1. Donasti = *donavisti*, and *nobis* = *mihi*

rus sub urbe] 'a country estate just outside the city'.

2. rus est mihi...] 'I have a larger estate in my window-box'. For window-boxes in *insulae* see the illustration on p. 15.

3. **Rus hoc dicere...**] 'can you call this an estate, style this an estate?'

5. 'Which the wing of a chirping grasshopper can cover.'

7. **Clausae cui...**] 'for which the petal of a rose-bud (lit. a closed rose) is a garland'.

8. **In quo...**] 'in which grass is not to be found any more than a leaf (for the perfumes) of Cosmus, or green pepper'. Cosmus was a celebrated perfumer.

10. **iacere rectus**] 'lie straight'.

11. 'A whole snake cannot find room to live.'

12. **male**] 'scarcely'.

13. **Consumpto moritur...**] 'a gnat dies (of starvation) when it has eaten up the willow-bed'.

14. **talpa est...**] 'a mole is my digger and ploughman'.

16. **Ridere**] 'split', lit. smile.

17. **colono**] Dative of agent: 'is feared by the farmer'.

18. **Tamquam sus Calydonius**] 'as though it were the Calydonian boar', which ravaged the countryside and was killed by Meleager after a famous hunt.

19. **sublata...**] 'carried off by the claw of a flying Procne (a swallow: Greek genitive), my whole crop is now in a swallow's nest'.

22. 'There isn't room even for half a Priapus', an image of whom often stood in gardens to scare birds.

23. **peracta**] 'when gathered in'.

24. **nuce picata**] 'in a nut-shell sealed with pitch': local ablative. *condimus* is 'we store'.

25. **Errasti** (=*erravisti*): 'you have made a mistake, but only in one letter'.

27. **Mallem dedisses**] A 'potential' wish: 'I should prefer that you had given me a *meal* when (*quo tempore*) you gave me a *mead*', i.e. a meadow; or 'a feed instead of a field'. *praedium* is an 'estate', *prandium* 'lunch'. Martial seems to be very ungrateful for the gift of this farm, small though it was, but there may have been reasons to justify his ingratitude.

XXXII. (XI, 53)

1. **cum sit edita...**] 'though she is sprung from the woad-stained Britons, how truly she has the feelings (*pectora*) of the Latin race!' *quam*, 'how', usually goes with an adjective or adverb.

3. **Quale decus**] Supply *habet*: 'what grace of form she has! Italian women (*matres*) might believe her to be (supply *eam*

esse) a Roman, Athenians might think her one of their own race'.

5. **Di bene**] Supply *fecerunt*: 'heaven has been kind to her, because she is prolific and has borne (children) to her faithful husband, and because she hopes while still young (*puella*) to see her sons- and daughters-in-law', i.e. see her children married.

quodque = *et quod.*

7. **Sic placeat...**] 'so may it be the will of heaven that...', lit. so may it please the gods that.... *superis* = *deis.*

8. **natis tribus**] Martial prays that she may have at least three children, to the parents of whom special privileges were given, the *ius trium liberorum.*

In IV, 13 Martial refers to the marriage of Claudia Peregrina ('the foreigner'), who is probably the Claudia Rufina of this poem, to his friend Pudens. St Paul includes Pudens and Claudia in a list of Christians in II Timothy iv, 21, but there is no evidence to connect the two pairs except the similarity of names.

XXXIII. (XII, 18)

2. **Subura**] It was the most crowded and least respectable district of Rome.

3. **collem...Dianae**] 'or tread the hill of queen Diana', the Aventine, where she had a temple.

4. **per limina potentiorum**] 'on the thresholds of the great', lit. more powerful men.

5. **te ventilat**] 'serves you as a fan', with one of its loose folds; or perhaps 'flaps round you'.

vagum...fatigant] 'the greater and the smaller Caelian hill wearies you as you wander about'. The Caelian hill was divided into two parts.

7. **repetita...mea...Bilbilis**] 'my Bilbilis, to which I have returned after many Decembers...has made me a rustic'.

9. **Auro et superba ferro**] 'proud of her gold and iron'. *superba* is slightly misplaced in this sentence.

10. **Hic pigri...**] 'here lazily, with pleasant toil, I cultivate...; such are the uncouth (comparative, lit. rather coarse) names for Celtiberian places'.

13. **Ingenti improboque somno**] 'sleep that is deep and shamefully long'.

14. **nec tertia hora**] 'not even the third hour of the day'. The time meant would depend on the seasons: the first hour began at dawn.

15. 'I am now making up to myself in full (*totum*) all my sleeplessness for thirty years', lit. whatever I had kept awake. *quidquid* is adverbial accusative. The final *-o* of *repono* is short.

17. **datur (mihi) petenti...]** 'the nearest clothes, lying on (lit. from) a broken chair, are given to me when I ask for them'.

19. **Surgentem (me)...]** 'when I get up, the (fire on the) hearth welcomes me, fed (*cultus*) with a glorious (lit. proud) pile of logs from the neighbouring oak-wood'.

21. **Multa** agrees with *olla*: 'which the bailiff's wife crowns with many a pot', for breakfast.

24. 'The close-cropped (*lēvis*, lit. smooth) bailiff gives the boy-slaves (their rations), and asks permission (supply *ut liceat*) for them to cut (lit. lay aside) their long hair', so that they can be ranked as grown-up slaves and do a full day's work.

26. **Sic me iuvat]** 'thus I delight (lit. it pleases me) to live, thus (I hope) to die', i.e. in the country.

PLINY THE YOUNGER

PUBLIUS CAECILIUS SECUNDUS, who afterwards took the name Plinius, was born in A.D. 61 or 62 at Novum Comum (Como) in Cisalpine Gaul, now part of northern Italy. His father, who was one of the officials of the town, died while Pliny was quite young, and the famous Verginius Rufus (see Pliny, IV, 1, note) was appointed his guardian, though Pliny was probably brought up by his mother Plinia and his uncle, the elder Pliny; the latter had held various high administrative posts open to *equites*, being at the time of his death in the eruption of Vesuvius in A.D. 79 commander of the fleet at Misenum; he was also an intimate friend of the emperor Vespasian and a student of boundless industry, and the author of many books on history, biography, languages, military science and natural history, of which only the last survives. The younger Pliny was educated at Comum, and studied rhetoric at Rome under Quintilian, and on the death of his uncle in 79 he inherited his large property and according to custom took his uncle's name, being henceforth called Gaius Plinius Caecilius Secundus; Plinius was his usual name, but some friends, notably the emperor Trajan, addressed him as Secundus. In the next year he began his legal career by pleading a cause in court, and also held a minor official post, and in 81 he served his time as military tribune, or junior officer, with the 3rd Gallic Legion in Syria. In 89 he was *quaestor* under Domitian (and *ex officio* a member of the senate), tribune in 91 and *praetor* in 93, in which year he managed to avoid being involved in the conspiracy and downfall of the Stoics, among whom he had many friends; he was soon after put in charge of the military treasury, although an accusation against him made by the informer Mettius Carus was found in the emperor's desk after his death in 96. Under Nerva (96–98) Pliny was in charge of the public treasury, and in 100, after successfully prosecuting the governor of Africa for extortion, in conjunction with the historian Tacitus, he was appointed *consul suffectus* (for two months of the year) by Trajan. His *Panegyric*, or address of thanks to the emperor for the honour, still exists. In 103 he was *augur*, a member of one of the priestly colleges, and in 105 president of the commission in charge of the bed and banks of the Tiber and the drainage of the city. Finally, in 111 he was sent out as governor of the province of Bithynia, in Asia Minor, which had recently been transferred from the senate to the emperor, and which required reforms in several directions. His letters to Trajan asking for advice and instruction on various matters, and the emperor's replies, are of great interest. Pliny probably died there in 113.

Pliny was a rich man all his life and possessed much property, including two villas on Lake Como, of which he was very fond, and he had a large number of slaves, whom he treated with unusual humanity. He did much for his native town both during his life and by bequests in his will, giving money for education, charities, and the building of public baths there.

He was three times married, his third wife, Calpurnia, bringing him great happiness and accompanying him to Bithynia. He had no children, but was granted the *ius trium liberorum* by Trajan. As an orator Pliny is very proud of his fame, which must have been genuinely great, though none of his speeches survive except his *Panegyric on Trajan*, which is of inferior quality. He practised mainly in the court of the *centumviri*, the 'Hundred' Court, which dealt with cases of inheritance and disputed property. But it is on his letters that his literary fame rests. He himself collected and published nine books of them, the tenth, his correspondence with Trajan, being published after his death; he took great pains to revise the first nine books before publication, and arranged the letters carefully so as to secure variety of interest, instead of keeping them in strict chronological order. The letters are written on many subjects connected with Pliny's everyday life and pursuits, such as public affairs, lawsuits, descriptions of country scenery and his villas, ghost stories, points of literary interest, and in particular an account of the eruption of Vesuvius in two letters to Tacitus. These, together with his report on Christianity in Bithynia and Trajan's reply, are the most famous and interesting in the collection. The tenth book is important as showing the relationship between a provincial governor and the emperor, who liked to be consulted on the smallest points of administration.

Pliny's letters are written in a polished style which is a happy blend of Ciceronian and Silver Latin. Variety of language, frequent omission of the verb 'to be' and much use of the historic infinitive in narrative are characteristic. They are clearly expressed, and each is well adapted to its subject, though they suffer in comparison with Cicero's Letters to his friends by being more artificial and dealing with subjects of less historical importance. Yet many of the letters are of the greatest interest and, like Martial's poems, surprisingly modern in outlook, though lacking in a sense of humour. Little is known of Pliny except from his letters, in which he is revealed as a man of considerable ability and kindliness, but of more than usual vanity, yet never jealous of the literary reputation of his friends, among whom he included Martial. An account of the relative social positions of Martial and Pliny is given on pp. 3 and 4.

PLINY

I. (Book III, 14) *The Death of Larcius Macedo*

To Acilius, describing how Macedo was attacked by his slaves while bathing, from the results of which he later died, and how the same man had once before had an unfortunate experience at the baths.

C. PLINIUS ACILIO SUO S.

Rem atrocem nec tantum epistula dignam Larcius 1
Macedo, vir praetorius, a servis suis passus est,
superbus alioqui dominus et saevus, et qui servisse
patrem suum parum, immo nimium meminisset.
Lavabatur in villa Formiana: repente eum servi 2
circumsistunt: alius fauces invadit, alius os verberat,
alius pectus et ventrem contundit; et, cum exanimem
putarent, abiciunt in fervens pavimentum, ut ex-
perirentur, an viveret. Ille, sive quia non sentiebat,
sive quia non sentire simulabat, inmobilis et extentus
fidem peractae mortis inplevit. Tum demum quasi 3
aestu solutus effertur; excipiunt servi fideliores,
feminae cum ululatu et clamore concurrunt. Ita et
vocibus excitatus et recreatus loci frigore sublatis
oculis agitatoque corpore vivere se (et iam tutum
erat) confitetur. Diffugiunt servi; quorum magna 4
pars comprehensa est, ceteri requiruntur. Ipse paucis
diebus aegre focilatus non sine ultionis solacio decessit
ita vivus vindicatus, ut occisi solent. Vides, quot 5
periculis, quot contumeliis, quot ludibriis simus
obnoxii; nec est, quod quisquam possit esse securus,
quia sit remissus et mitis; non enim iudicio domini,
sed scelere perimuntur. Verum haec hactenus.

6 Quid praeterea novi? quid? nihil; alioqui sub-
iungerem: nam et charta adhuc superest, et dies
feriatus patitur plura contexi. Addam, quod oppor-
tune de eodem Macedone succurrit. Cum in publico
Romae lavaretur, notabilis atque etiam, ut exitus
7 docuit, ominosa res accidit. Eques Romanus a servo
eius, ut transitum daret, manu leviter admonitus
convertit se nec servum, a quo erat tactus, sed ipsum
Macedonem tam graviter palma percussit, ut paene
8 concideret. Ita balineum illi quasi per gradus quos-
dam primum contumeliae locus, deinde exitii fuit.
Vale.

II. (III, 16) *The True Greatness of the elder Arria*

To P. Metilius Sabinus Nepos, later a provincial governor, giving an account of Arria's fortitude in concealing her son's death from her sick husband, and her determination to accompany her husband to Rome and die with him by any means, when he was accused of treason; such actions were perhaps more praiseworthy even than her famous suicide.

C. PLINIUS NEPOTI SUO S.

1 Adnotasse videor facta dictaque virorum feminarum-
que alia clariora esse, alia maiora. Confirmata est
2 opinio mea hesterno Fanniae sermone. Neptis haec
Arriae illius, quae marito et solacium mortis et ex-
emplum fuit. Multa referebat aviae suae non minora
hoc, sed obscuriora; quae tibi existimo tam mirabilia
3 legenti fore, quam mihi audienti fuerunt. Aegrotabat
Caecina Paetus, maritus eius, aegrotabat et filius,
uterque mortifere, ut videbatur. Filius decessit exi-
mia pulchritudine, pari verecundia et parentibus non
4 minus ob alia carus, quam quod filius erat. Huic illa
ita funus paravit, ita duxit exequias, ut ignoraret
maritus; quin immo, quotiens cubiculum eius intra-

ret, vivere filium atque etiam commodiorem esse
simulabat ac persaepe interroganti, quid ageret puer,
respondebat: 'Bene quievit, libenter cibum sumpsit'.
Deinde, cum diu cohibitae lacrimae vincerent pro- 5
rumperentque, egrediebatur: tum se dolori dabat;
satiata siccis oculis, composito vultu redibat, tam-
quam orbitatem foris reliquisset. Praeclarum quidem 6
illud eiusdem, ferrum stringere, perfodere pectus,
extrahere pugionem, porrigere marito, addere vocem
inmortalem ac paene divinam: 'Paete, non dolet'.
Sed tamen ista facienti, dicenti gloria et aeternitas
ante oculos erant; quo maius est sine praemio
aeternitatis, sine praemio gloriae abdere lacrimas,
operire luctum amissoque filio matrem adhuc agere.
Scribonianus arma in Illyrico contra Claudium move- 7
rat; fuerat Paetus in partibus, occiso Scriboniano
Romam trahebatur. Erat ascensurus navem. Arria 8
milites orabat, ut simul inponeretur. 'Nempe enim',
inquit, 'daturi estis consulari viro servulos aliquos,
quorum e manu cibum capiat, a quibus vestiatur, a
quibus calcietur; omnia sola praestabo.' Non im- 9
petravit: conduxit piscatoriam naviculam ingensque
navigium minimo secuta est. Eadem apud Claudium
uxori Scriboniani, cum illa profiteretur indicium,
'Ego', inquit, 'te audiam, cuius in gremio Scriboni-
anus occisus est, et vivis?' Ex quo manifestum est
ei consilium pulcherrimae mortis non subitum fuisse.
Quin etiam, cum Thrasea, gener eius, deprecaretur, 10
ne mori pergeret, interque alia dixisset: 'Vis ergo
filiam tuam, si mihi pereundum fuerit, mori mecum?'
respondit: 'Si tam diu tantaque concordia vixerit
tecum quam ego cum Paeto, volo.' Auxerat hoc re- 11

sponso curam suorum, attentius custodiebatur: sensit et 'Nihil agitis' inquit; 'potestis enim efficere, ut male
12 moriar, ut non moriar, non potestis.' Dum haec dicit, exiluit cathedra adversoque parieti caput ingenti impetu inpegit et corruit. Focilata 'Dixeram' inquit 'vobis inventuram me quamlibet duram ad mortem
13 viam, si vos facilem negassetis.' Videnturne haec tibi maiora illo 'Paete, non dolet', ad quod per haec perventum est? cum interim illud quidem ingens fama, haec nulla circumfert. Unde colligitur, quod initio dixi, alia esse clariora, alia maiora. Vale.

III. (III, 21) *The Death of Martial*

To Cornelius Priscus, an ex-consul and afterwards governor of Asia, expressing sorrow at the news of Martial's death. Pliny gives a brief sketch of the poet's character, and quotes some verses addressed to himself.

C. PLINIUS CORNELIO PRISCO SUO S.

1 Audio Valerium Martialem decessisse et moleste fero. Erat homo ingeniosus, acutus, acer, et qui plurimum in scribendo et salis haberet et fellis nec candoris
2 minus. Prosecutus eram viatico secedentem; dederam hoc amicitiae, dederam etiam versiculis, quos de
3 me composuit. Fuit moris antiqui eos, qui vel singulorum laudes vel urbium scripserant, aut honoribus aut pecunia ornare; nostris vero temporibus ut alia speciosa et egregia ita hoc inprimis exolevit. Nam, postquam desiimus facere laudanda, laudari quoque
4 ineptum putamus. Quaeris, qui sint versiculi, quibus gratiam rettuli. Remitterem te ad ipsum volumen, nisi quosdam tenerem; tu, si placuerint hi, ceteros in

MAP OF
CENTRAL ITALY
from
ROME to NAPLES
Scale of English Miles
0 5 10 15 30
Via Flaminia
Via Salaria
Saxa Rubra
Fidenae
Via Tiburtina
Tibur
Alsium
ROMA
Praeneste
Tusculum
Alba Longa
Albanum
Ostia
LATIUM
Setia
Via Appia
Formiae
Circei
Caieta
CAMPANIA
NEAPOLIS
M. VESUVIUS
Baiae
Misenum Pr.
Pompeii
Stabiae
CAPREAE I.
MARE
TYRRHENUM

5 libro requires. Adloquitur Musam, mandat, ut domum meam Esquiliis quaerat, adeat reverenter:

Sed ne tempore non tuo disertam
pulses ebria ianuam, videto.
Totos dat tetricae dies Minervae,
dum centum studet auribus virorum
hoc, quod saecula posterique possint
Arpinis quoque comparare chartis.
Seras tutior ibis ad lucernas:
haec hora est tua, cum furit Lyaeus,
cum regnat rosa, cum madent capilli.
Tunc me vel rigidi legant Catones.

6 Meritone eum, qui haec de me scripsit, et tunc dimisi amicissime et nunc ut amicissimum defunctum esse doleo? Dedit enim mihi, quantum maximum potuit, daturus amplius, si potuisset. Tametsi quid homini potest dari maius quam gloria et laus et aeternitas harum? At non erunt aeterna, quae scripsit: non erunt fortasse, ille tamen scripsit, tamquam essent futura. Vale.

IV. (VI, 10) *The neglected Tomb of Verginius Rufus*

To Lucceius Albinus, a barrister who was associated with Pliny in several trials, complaining that the tomb of even such a man as Verginius lay neglected soon after his death, although Verginius had ordered an epitaph commemorating the great event of his career to be inscribed on it.

C. PLINIUS ALBINO SUO S.

1 Cum venissem in socrus meae villam Alsiensem, quae aliquando Rufi Vergini fuit, ipse mihi locus optimi illius et maximi viri desiderium non sine dolore renovavit. Hunc enim incolere secessum atque etiam

senectutis suae nidulum vocare consueverat. Quo- 2
cunque me contulissem, illum animus, illum oculi
requirebant. Libuit etiam monimentum eius videre,
et vidisse paenituit. Est enim adhuc inperfectum, 3
nec difficultas operis in causa, modici ac potius exigui,
sed inertia eius, cui cura mandata est. Subit indig-
natio cum miseratione post decimum mortis annum
reliquias neglectumque cinerem sine titulo, sine
nomine iacere, cuius memoria orbem terrarum gloria
pervagetur. At ille mandaverat caveratque, ut di- 4
vinum illud et inmortale factum versibus inscriberetur:

Hic situs est Rufus, pulso qui Vindice quondam
imperium adseruit non sibi, sed patriae.

Tam rara in amicitiis fides, tam parata oblivio mortu- 5
orum, ut ipsi nobis debeamus etiam conditoria extruere
omniaque heredum officia praesumere. Nam cui non
est verendum, quod videmus accidisse Verginio?
cuius iniuriam ut indigniorem sic etiam notiorem
ipsius claritas facit. Vale.

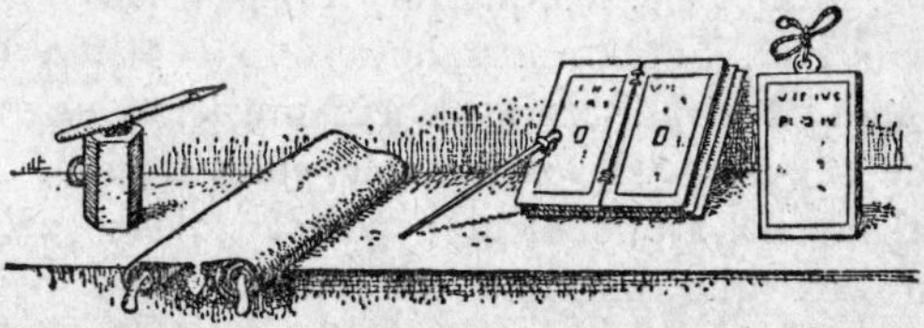

Writing materials. (Reed-pen and inkpot; papyrus roll; stilus and wax tablets; calculation table.)

V. (VI, 16) *The Eruption of Vesuvius and Death of the Elder Pliny*

To Tacitus, the great historian, who wanted an account of the eruption for his history. 1–10. On 24 August A.D. 79, the elder Pliny determined to inspect from close at hand a remarkable cloud which was rising from Vesuvius. His scientific interest and heroic desire to save the lives of others led him straight towards the eruption. 11–22. In spite of the danger, Pliny then sailed to Stabiae to save another friend, but after behaving with great coolness in constant peril from falling houses, he was prevented by the rough sea from returning, and was choked by the fumes of the volcano and died peacefully.

C. PLINIUS TACITO SUO S.

1 Petis, ut tibi avunculi mei exitum scribam, quo verius tradere posteris possis. Gratias ago; nam video morti eius, si celebretur a te, inmortalem
2 gloriam esse propositam. Quamvis enim pulcherrimarum clade terrarum, et populi et urbis memorabili casu, quasi semper victurus occiderit, quamvis ipse plurima opera et mansura condiderit, multum tamen perpetuitati eius scriptorum tuorum aeternitas addet.
3 Equidem beatos puto, quibus deorum munere datum est aut facere scribenda aut scribere legenda, beatissimos vero, quibus utrumque. Horum in numero avunculus meus et suis libris et tuis erit. Quo libentius suscipio, deposco etiam, quod iniungis.

4 Erat Miseni classemque imperio praesens regebat. Nonum Kal. Septembres hora fere septima mater mea indicat ei adparere nubem inusitata et magni-
5 tudine et specie. Usus ille sole, mox frigida, gustaverat iacens studebatque; poscit soleas, ascendit locum, ex quo maxime miraculum illud conspici poterat. Nubes, incertum procul intuentibus, ex quo monte (Vesuvium fuisse postea cognitum est), oriebatur, cuius similitudinem et formam non alia magis
6 arbor quam pinus expresserit. Nam longissimo velut trunco elata in altum quibusdam ramis diffundebatur, credo, quia recenti spiritu evecta, dein senescente eo destituta aut etiam pondere suo victa in latitudinem

vanescebat, candida interdum, interdum sordida et
maculosa, prout terram cineremve sustulerat.
Magnum propiusque noscendum ut eruditissimo viro 7
visum. Iubet Liburnicam aptari: mihi, si venire una
vellem, facit copiam; respondi studere me malle, et
forte ipse, quod scriberem, dederat. Egrediebatur 8
domo: accipit codicillos Rectinae Tasci imminenti
periculo exterritae (nam villa eius subiacebat, nec
ulla nisi navibus fuga); ut se tanto discrimini eriperet,
orabat. Vertit ille consilium et, quod studioso animo 9
incohaverat, obit maximo. Deducit quadriremes,
ascendit ipse non Rectinae modo, sed multis (erat
enim frequens amoenitas orae) laturus auxilium.
Properat illuc, unde alii fugiunt, rectumque cursum, 10
recta gubernacula in periculum tenet adeo solutus
metu, ut omnes illius mali motus, omnes figuras, ut
deprehenderat oculis, dictaret enotaretque. Iam 11
navibus cinis incidebat, quo propius accederent,
calidior et densior, iam pumices etiam nigrique et
ambusti et fracti igne lapides, iam vadum subitum
ruinaque montis litora obstantia. Cunctatus paulum,
an retro flecteret, mox gubernatori, ut ita faceret,
monenti 'Fortes', inquit, 'fortuna iuvat: Pomponi-
anum pete'. Stabiis erat diremptus sinu medio (nam 12
sensim circumactis curvatisque litoribus mare in-
funditur); ibi quamquam nondum periculo adpropin-
quante, conspicuo tamen et, cum cresceret, proximo
sarcinas contulerat in naves certus fugae, si contra-
rius ventus resedisset; quo tunc avunculus meus
secundissimo invectus complectitur trepidantem, con-
solatur, hortatur, utque timorem eius sua securitate
leniret, deferri in balineum iubet: lotus accubat, cenat

aut hilaris aut, quod est aeque magnum, similis hilari.
13 Interim e Vesuvio monte pluribus locis latissimae
flammae altaque incendia relucebant, quorum fulgor
et claritas tenebris noctis excitabatur. Ille agrestium
trepidatione ignes relictos desertasque villas per soli-
tudinem ardere in remedium formidinis dictitabat.
Tum se quieti dedit et quievit verissimo quidem
somno. Nam meatus animae, qui illi propter ampli-
tudinem corporis gravior et sonantior erat, ab iis, qui
14 limini obversabantur, audiebatur. Sed area, ex qua
diaeta adibatur, ita iam cinere mixtisque pumicibus
oppleta surrexerat ut, si longior in cubiculo mora,
15 exitus negaretur. Excitatus procedit seque Pom-
poniano ceterisque, qui pervigilaverant, reddit. In
commune consultant, intra tecta subsistant an in
aperto vagentur. Nam crebris vastisque tremoribus
tecta nutabant et quasi emota sedibus suis nunc huc,
16 nunc illuc abire aut referri videbantur. Sub dio rursus
quamquam levium exesorumque pumicum casus
metuebatur; quod tamen periculorum collatio elegit.
Et apud illum quidem ratio rationem, apud alios
timorem timor vicit. Cervicalia capitibus inposita
linteis constringunt: id munimentum adversus inci-
17 dentia fuit. Iam dies alibi, illic nox omnibus noctibus
nigrior densiorque; quam tamen faces multae varia-
que lumina solabantur. Placuit egredi in litus et ex
proximo adspicere, ecquid iam mare admitteret; quod
18 adhuc vastum et adversum permanebat. Ibi super
abiectum linteum recubans semel atque iterum frigi-
dam poposcit hausitque. Deinde flammae flammar-
umque praenuntius odor sulpuris alios in fugam
19 vertunt, excitant illum. Innixus servulis duobus ad-

surrexit et statim concidit, ut ego colligo, crassiore
caligine spiritu obstructo clausoque stomacho, qui illi
natura invalidus et angustus et frequenter inter-
aestuans erat. Ubi dies redditus (is ab eo, quem 20
novissime viderat, tertius), corpus inventum inte-
grum, inlaesum opertumque, ut fuerat indutus:
habitus corporis quiescenti quam defuncto similior.

Interim Miseni ego et mater. Sed nihil ad historiam, 21
nec tu aliud quam de exitu eius scire voluisti. Finem
ergo faciam. Unum adiciam, omnia me, quibus inter- 22
fueram, quaeque statim, cum maxime vera memo-
rantur, audieram, persecutum. Tu potissima excerpes;
aliud est enim epistulam, aliud historiam, aliud
amico, aliud omnibus scribere. Vale.

VI. (VI, 20) *The Younger Pliny's experiences during the Eruption of Vesuvius*

To Tacitus, in reply to a request that Pliny should describe his own adventures. 1–11. Pliny and his mother were at Misenum, when an earthquake took place, accompanied by lightning and the sucking back of the sea. In spite of appeals by a friend, they refused to leave before hearing of the elder Pliny's safety. 12–20. Finally Pliny persuaded his mother to set out, in complete darkness, while cries of terror could be heard everywhere. Flames approached, but receded again, and dawn revealed a thick layer of ash covering everything. They returned to Misenum to await news of the elder Pliny, though the earth-tremors still continued.

C. PLINIUS TACITO SUO S.

Ais te adductum litteris, quas exigenti tibi de morte 1
avunculi mei scripsi, cupere cognoscere, quos ego
Miseni relictus (id enim ingressus abruperam) non
solum metus, verum etiam casus pertulerim: 'quam-
quam animus meminisse horret, incipiam'. Profecto 2
avunculo ipse reliquum tempus studiis (ideo enim

remanseram) impendi; mox balineum, cena, somnus
3 inquietus et brevis. Praecesserat per multos dies
tremor terrae minus formidolosus, quia Campaniae
solitus; illa vero nocte ita invaluit, ut non moveri
4 omnia, sed verti crederentur. Inrumpit cubiculum
meum mater: surgebam in vicem, si quiesceret, ex-
citaturus. Resedimus in area domus, quae mare a
5 tectis modico spatio dividebat. Dubito, constantiam
vocare an imprudentiam debeam (agebam enim duo-
devicensimum annum): posco librum Titi Livi et
quasi per otium lego atque etiam, ut coeperam, ex-
cerpo. Ecce amicus avunculi, qui nuper ad eum ex
Hispania venerat, ut me et matrem sedentes, me vero
etiam legentem videt, illius patientiam, securitatem
meam corripit. Nihilo segnius ego intentus in librum.
6 Iam hora diei prima, et adhuc dubius et quasi langui-
dus dies. Iam quassatis circumiacentibus tectis,
quamquam in aperto loco, angusto tamen, magnus et
7 certus ruinae metus. Tum demum excedere oppido
visum: sequitur vulgus attonitum, quodque in pavore
simile prudentiae, alienum consilium suo praefert
ingentique agmine abeuntes premit et inpellit. Egressi
8 tecta consistimus. Multa ibi miranda, multas for-
midines patimur. Nam vehicula, quae produci iusse-
ramus, quamquam in planissimo campo, in contrarias
partes agebantur ac ne lapidibus quidem fulta in
9 eodem vestigio quiescebant. Praeterea mare in se
resorberi et tremore terrae quasi repelli videbamus.
Certe processerat litus multaque animalia maris siccis
harenis detinebat. Ab altero latere nubes atra et
horrenda ignei spiritus tortis vibratisque discursibus
rupta in longas flammarum figuras dehiscebat: fulgo-

ribus illae et similes et maiores erant. Tum vero idem 10
ille ex Hispania amicus acrius et instantius 'Si frater'
inquit, 'tuus, tuus avunculus vivit, vult esse vos
salvos; si periit, superstites voluit: proinde quid ces-
satis evadere?' Respondimus non commissuros nos,
ut de salute illius incerti nostrae consuleremus. Non 11
moratus ultra proripit se effusoque cursu periculo
aufertur. Nec multo post illa nubes descendere in
terras, operire maria; cinxerat Capreas et absconderat,
Miseni quod procurrit, abstulerat. Tum mater orare, 12
hortari, iubere, quoquo modo fugerem; posse enim
iuvenem, se et annis et corpore gravem bene mori-
turam, si mihi causa mortis non fuisset. Ego contra
salvum me nisi una non futurum; deinde manum eius
amplexus addere gradum cogo. Paret aegre incu-
satque se, quod me moretur. Iam cinis, adhuc tamen 13
rarus. Respicio: densa caligo tergis imminebat, quae
nos torrentis modo infusa terrae sequebatur. 'De-
flectamus', inquam, 'dum videmus, ne in via strati
comitantium turba in tenebris obteramur.' Vix con- 14
sederamus, et nox, non qualis inlunis aut nubila, sed
qualis in locis clausis lumine extincto. Audires ulu-
latus feminarum, infantium quiritatus, clamores
virorum; alii parentes, alii liberos, alii coniuges voci-
bus requirebant, vocibus noscitabant; hi suum casum,
illi suorum miserabantur; erant qui metu mortis
mortem precarentur; multi ad deos manus tollere, 15
plures nusquam iam deos ullos, aeternamque illam et
novissimam noctem mundo interpretabantur. Nec
defuerunt, qui fictis mentitisque terroribus vera peri-
cula augerent. Aderant, qui Miseni illud ruisse, illud
ardere falso, sed credentibus nuntiabant. Paulum 16

reluxit; quod non dies nobis, sed adventantis ignis
indicium videbatur. Et ignis quidem longius substitit,
tenebrae rursus, cinis rursus multus et gravis. Hunc
identidem adsurgentes excutiebamus; operti alioquin
17 atque etiam oblisi pondere essemus. Possem gloriari
non gemitum mihi, non vocem parum fortem in tantis
periculis excidisse, nisi me cum omnibus, omnia
mecum perire misero, magno tamen mortalitatis
18 solacio credidissem. Tandem illa caligo tenuata quasi
in fumum nebulamve discessit; mox dies verus, sol
etiam effulsit, luridus tamen, qualis esse, cum deficit,
solet. Occursabant trepidantibus adhuc oculis mutata
19 omnia altoque cinere tamquam nive obducta. Re-
gressi Misenum curatis utcunque corporibus sus-
pensam dubiamque noctem spe ac metu exegimus.
Metus praevalebat; nam et tremor terrae persevera-
bat, et plerique lymphati terrificis vaticinationibus et
20 sua et aliena mala ludificabantur. Nobis tamen ne
tunc quidem, quamquam et expertis periculum et
expectantibus, abeundi consilium, donec de avunculo
nuntius.

Haec nequaquam historia digna non scripturus leges et tibi, scilicet qui requisisti, inputabis, si digna ne epistula quidem videbuntur. Vale.

VII. (VII, 27) *Some Ghost Stories*

To Licinius Sura, a Spaniard, who was a distinguished soldier and a friend of Nerva, Trajan and Hadrian. Three ghost stories: (i) the phantom that foretold the promotion and death of Curtius Rufus; (ii) the haunted house at Athens, where the ghost of a dead man was laid by a certain philosopher; (iii) the cutting short of the hair of two of Pliny's slaves in the night.

C. PLINIUS SURAE SUO S.

Et mihi discendi et tibi docendi facultatem otium 1
praebet. Igitur perquam velim scire, esse phantas-
mata et habere propriam figuram numenque aliquod
putes an inania et vana ex metu nostro imaginem
accipere. Ego ut esse credam, inprimis eo ducor, quod 2
audio accidisse Curtio Rufo. Tenuis adhuc et obscurus
obtinenti Africam comes haeserat. Inclinato die
spatiabatur in porticu: offertur ei mulieris figura
humana grandior pulchriorque; perterrito Africam se
futurorum praenuntiam dixit; iturum enim Romam
honoresque gesturum atque etiam cum summo im-
perio in eandem provinciam reversurum ibique
moriturum. Facta sunt omnia. Praeterea accedenti 3
Carthaginem egredientique nave eadem figura in
litore occurrisse narratur. Ipse certe implicitus morbo
futura praeteritis, adversa secundis auguratus spem
salutis nullo suorum desperante proiecit.

Iam illud nonne et magis terribile et non minus 4
mirum est, quod exponam, ut accepi? Erat Athenis 5
spatiosa et capax domus, sed infamis et pestilens. Per
silentium noctis sonus ferri et, si attenderes acrius,
strepitus vinculorum longius primo, deinde e proximo
reddebatur: mox adparebat idolon, senex macie et
squalore confectus, promissa barba, horrenti capillo;
cruribus compedes, manibus catenas gerebat quatie-
batque. Inde inhabitantibus tristes diraeque noctes 6
per metum vigilabantur; vigiliam morbus et crescente
formidine mors sequebatur. Nam interdiu quoque,
quamquam abscesserat imago, memoria imaginis
oculis inerrabat, longiorque causis timoris timor erat.

Deserta inde et damnata solitudine domus totaque
illi monstro relicta; proscribebatur tamen, seu quis
emere seu quis conducere ignarus tanti mali vellet.
7 Venit Athenas philosophus Athenodorus, legit titu-
lum auditoque pretio, quia suspecta vilitas, percunc-
tatus omnia docetur ac nihilo minus, immo tanto
magis conducit. Ubi coepit advesperascere, iubet
sterni sibi in prima domus parte, poscit pugillares,
stilum, lumen: suos omnes in interiora dimittit, ipse
ad scribendum animum, oculos, manum intendit, ne
vacua mens audita simulacra et inanes sibi metus
8 fingeret. Initio, quale ubique, silentium noctis, dein
concuti ferrum, vincula moveri. Ille non tollere oculos,
non remittere stilum, sed offirmare animum auribus-
que praetendere. Tum crebrescere fragor, adventare
et iam ut in limine, iam ut intra limen audiri. Respicit,
9 videt agnoscitque narratam sibi effigiem. Stabat in-
nuebatque digito similis vocanti. Hic contra, ut
paulum expectaret, manu significat rursusque ceris et
stilo incumbit. Illa scribentis capiti catenis insonabat.
Respicit rursus idem quod prius innuentem nec
10 moratus tollit lumen et sequitur. Ibat illa lento gradu
quasi gravis vinculis. Postquam deflexit in aream
domus, repente dilapsa deserit comitem. Desertus
11 herbas et folia concerpta signum loco ponit. Postero
die adit magistratus, monet, ut illum locum effodi
iubeant. Inveniuntur ossa inserta catenis et im-
plicita, quae corpus aevo terraque putrefactum nuda
et exesa reliquerat vinculis: collecta publice sepe-
12 liuntur. Domus postea rite conditis manibus caruit.
Et haec quidem adfirmantibus credo; illud adfirmare
aliis possum.

A Roman temple

Est libertus mihi non inlitteratus. Cum hoc minor
frater eodem lecto quiescebat. Is visus est sibi cernere
quendam in toro residentem admoventemque capiti
suo cultros atque etiam ex ipso vertice amputantem
capillos. Ubi inluxit, ipse circa verticem tonsus,
13 capilli iacentes reperiuntur. Exiguum temporis
medium, et rursus simile aliud priori fidem fecit. Puer
in paedagogio mixtus pluribus dormiebat: venerunt
per fenestras (ita narrat) in tunicis albis duo cuban-
temque detonderunt et, qua venerant, recesserunt.
Hunc quoque tonsum sparsosque circa capillos dies
14 ostendit. Nihil notabile secutum, nisi forte quod non
fui reus, futurus, si Domitianus, sub quo haec acci-
derunt, diutius vixisset. Nam in scrinio eius datus a
Caro de me libellus inventus est; ex quo coniectari
potest, quia reis moris est summittere capillum, re-
cisos meorum capillos depulsi, quod imminebat, peri-
culi signum fuisse.

15 Proinde rogo, eruditionem tuam intendas. Digna
res est, quam diu multumque consideres, ne ego qui-
16 dem indignus, cui copiam scientiae tuae facias. Licet
etiam utramque in partem, ut soles, disputes, ex
altera tamen fortius, ne me suspensum incertumque
dimittas, cum mihi consulendi causa fuerit, ut dubi-
tare desinerem. Vale.

VIII. (IX, 23) *Pliny's Literary Fame*

To Maximus, formerly quaestor in Bithynia, recounting two anecdotes about Pliny's literary reputation.

C. PLINIUS MAXIMO SUO S.

1 Frequenter agenti mihi evenit, ut centumviri, cum
diu se intra iudicum auctoritatem gravitatemque
tenuissent, omnes repente quasi victi coactique con-

surgerent laudarentque; frequenter e senatu famam, 2
qualem maxime optaveram, rettuli: numquam tamen maiorem cepi voluptatem quam nuper ex sermone Corneli Taciti. Narrabat sedisse secum circensibus proximis equitem Romanum. Hunc post varios eruditosque sermones requisisse: 'Italicus es an provincialis?' Se respondisse: 'Nosti me et quidem ex
studiis.' Ad hoc illum: 'Tacitus es an Plinius?' 3
Exprimere non possum, quam sit iucundum mihi, quod nomina nostra quasi litterarum propria, non hominum, litteris redduntur, quod uterque nostrum his etiam ex studiis notus, quibus aliter ignotus est.

Accidit aliud ante pauculos dies simile. Recumbe- 4
bat mecum vir egregius, Fadius Rufinus, super eum municeps ipsius, qui illo die primum venerat in urbem; cui Rufinus demonstrans me: 'Vides hunc?' Multa deinde de studiis nostris, et ille: 'Plinius est'
inquit. Verum fatebor, capio magnum laboris mei 5
fructum. An, si Demosthenes iure laetatus est, quod illum anus Attica ita noscitavit: *Οὗτός ἐστι Δημοσθένης*, ego celebritate nominis mei gaudere non debeo? Ego vero et gaudeo et gaudere me dico.
Neque enim vereor, ne iactantior videar, cum de me 6
aliorum iudicium, non meum profero, praesertim apud te, qui nec ullius invides laudibus et faves nostris. Vale.

IX. (IX, 33) *The Boy and the Dolphin*

To Caninius Rufus, a poet and a man of wealth, relating the true story of a dolphin who used to take a boy on its back, and even come out of the water to bask on the shore, until it was frightened off by a foolish official, and finally killed by the local magistrates to prevent crowds from flocking to their town.

C. PLINIUS CANINIO SUO S.

1 Incidi in materiam veram, sed simillimam fictae dig-
namque isto laetissimo, altissimo planeque poëtico
ingenio, incidi autem, dum super cenam varia mira-
cula hinc inde referuntur. Magna auctori fides:
tametsi quid poëtae cum fide? Is tamen auctor, cui
2 bene vel historiam scripturus credidisses. Est in
Africa Hipponensis colonia mari proxima. Adiacet
navigabile stagnum: ex hoc in modum fluminis
aestuarium emergit, quod vice alterna, prout aestus
aut repressit aut inpulit, nunc infertur mari, nunc
3 redditur stagno. Omnis hic aetas piscandi, navigandi
atque etiam natandi studio tenetur, maxime pueri,
quos otium lususque sollicitat. His gloria et virtus
altissime provehi: victor ille, qui longissime ut litus
4 ita simul natantes reliquit. Hoc certamine puer qui-
dam audentior ceteris in ulteriora tendebat. Delphinus
occurrit, et nunc praecedere puerum, nunc sequi,
nunc circumire, postremo subire, deponere, iterum
subire trepidantemque perferre primum in altum,
mox flectit ad litus redditque terrae et aequalibus.
5 Serpit per coloniam fama: concurrere omnes, ipsum
puerum tamquam miraculum adspicere, interrogare,
audire, narrare. Postero die obsident litus, prospectant
mare et si quid est mari simile. Natant pueri, inter hos
ille, sed cautius. Delphinus rursus ad tempus, rursus
ad puerum. Fugit ille cum ceteris. Delphinus, quasi
invitet, revocet, exilit, mergitur variosque orbes im-
6 plicat expeditque. Hoc altero die, hoc tertio, hoc
pluribus, donec homines innutritos mari subiret
timendi pudor. Accedunt et adludunt et appellant,
tangunt etiam pertrectantque praebentem. Crescit

audacia experimento. Maxime puer, qui primus ex-
pertus est, adnatat nanti, insilit tergo, fertur refertur-
que, agnosci se, amari putat, amat ipse; neuter timet,
neuter timetur; huius fiducia, mansuetudo illius
augetur. Nec non alii pueri dextra laevaque simul 7
eunt hortantes monentesque. Ibat una (id quoque
mirum) delphinus alius, tantum spectator et comes.
Nihil enim simile aut faciebat aut patiebatur, sed
alterum illum ducebat, reducebat ut puerum ceteri
pueri. Incredibile, tam verum tamen quam priora, 8
delphinum gestatorem collusoremque puerorum in
terram quoque extrahi solitum harenisque siccatum,
ubi incaluisset, in mare revolvi. Constat Octavium 9
Avitum, legatum proconsulis, in litus educto religione
prava superfudisse unguentum, cuius illum novi-
tatem odoremque in altum refugisse nec nisi post
multos dies visum languidum et maestum, mox red-
ditis viribus priorem lasciviam et solita ministeria
repetisse. Confluebant omnes ad spectaculum magi- 10
stratus, quorum adventu et mora modica res publica
novis sumptibus atterebatur. Postremo locus ipse
quietem suam secretumque perdebat: placuit occulte
interfici, ad quod coibatur. Haec tu qua miseratione, 11
qua copia deflebis, ornabis, attolles! Quamquam non
est opus adfingas aliquid aut adstruas; sufficit, ne ea,
quae sunt vera, minuantur. Vale.

X. (X, 96) *The Christians in Bithynia*

To the emperor Trajan, written when Pliny was governor of Bithynia, asking for advice how to deal with the Christians; he had executed obstinate ones, and dismissed those who recanted and worshipped the emperor's image. Torture could extract no admission of guilt, and an opportunity for repentance might induce some of the rapidly increasing sect to abandon their superstition.

C. PLINIUS TRAIANO IMPERATORI

1 Sollemne est mihi, domine, omnia, de quibus dubito, ad te referre. Quis enim potest melius vel cunctationem meam regere vel ignorantiam instruere? Cognitionibus de Christianis interfui numquam: ideo nescio, quid et quatenus aut puniri soleat aut quaeri.
2 Nec mediocriter haesitavi, sitne aliquod discrimen aetatum, an quamlibet teneri nihil a robustioribus differant, detur paenitentiae venia, an ei, qui omnino Christianus fuit, desisse non prosit, nomen ipsum, si flagitiis careat, an flagitia cohaerentia nomini puniantur. Interim in iis, qui ad me tamquam Christiani
3 deferebantur, hunc sum secutus modum. Interrogavi ipsos, an essent Christiani. Confitentes iterum ac tertio interrogavi supplicium minatus: perseverantes duci iussi. Neque enim dubitabam, qualecunque esset, quod faterentur, pertinaciam certe et inflexibilem
4 obstinationem debere puniri. Fuerunt alii similis amentiae, quos, quia cives Romani erant, adnotavi in urbem remittendos. Mox ipso tractatu, ut fieri solet, diffundente se crimine plures species inciderunt.
5 Propositus est libellus sine auctore multorum nomina continens. Qui negabant esse se Christianos aut fuisse, cum praeeunte me deos appellarent et imagini tuae, quam propter hoc iusseram cum simulacris numinum afferri, ture ac vino supplicarent, praeterea male dicerent Christo, quorum nihil posse cogi dicuntur, qui sunt re vera Christiani, dimittendos esse
6 putavi. Alii ab indice nominati esse se Christianos dixerunt et mox negaverunt; fuisse quidem, sed desisse, quidam ante triennium, quidam ante plures

annos, non nemo etiam ante viginti. Hi quoque 7
omnes et imaginem tuam deorumque simulacra vene-
rati sunt et Christo male dixerunt. Adfirmabant
autem hanc fuisse summam vel culpae suae vel
erroris, quod essent soliti stato die ante lucem con-
venire carmenque Christo quasi deo dicere secum in
vicem seque sacramento non in scelus aliquod ob-
stringere, sed ne furta, ne latrocinia, ne adulteria
committerent, ne fidem fallerent, ne depositum ap-
pellati abnegarent. Quibus peractis morem sibi
discedendi fuisse rursusque coeundi ad capiendum
cibum, promiscuum tamen et innoxium; quod ipsum
facere desisse post edictum meum, quo secundum
mandata tua hetaerias esse vetueram. Quo magis 8
necessarium credidi ex duabus ancillis, quae ministrae
dicebantur, quid esset veri, et per tormenta quaerere.
Nihil aliud inveni quam superstitionem pravam, in-
modicam. Ideo dilata cognitione ad consulendum te
decucurri. Visa est enim mihi res digna consultatione, 9
maxime propter periclitantium numerum. Multi
enim omnis aetatis, omnis ordinis, utriusque sexus
etiam vocantur in periculum et vocabuntur. Neque
civitates tantum, sed vicos etiam atque agros super- 10
stitionis istius contagio pervagata est; quae videtur
sisti et corrigi posse. Certe satis constat prope iam
desolata templa coepisse celebrari, et sacra sollemnia
diu intermissa repeti pastumque venire victimarum,
cuius adhuc rarissimus emptor inveniebatur. Ex quo
facile est opinari, quae turba hominum emendari
possit, si sit paenitentiae locus.

XI. (X, 97) *Trajan's Reply about the Christians*

Pliny had acted rightly, but search for Christians must not be made; they should be pardoned if they worshipped Roman gods, and anonymous accusations must be ignored.

TRAIANUS PLINIO

1 Actum, quem debuisti, mi Secunde, in excutiendis causis eorum, qui Christiani ad te delati fuerant, secutus es. Neque enim in universum aliquid, quod
2 quasi certam formam habeat, constitui potest. Conquirendi non sunt; si deferantur et arguantur, puniendi sunt, ita tamen, ut, qui negaverit se Christianum esse idque re ipsa manifestum fecerit, id est supplicando dis nostris, quamvis suspectus in praeteritum, veniam ex paenitentia impetret. Sine auctore vero propositi libelli in nullo crimine locum habere debent. Nam et pessimi exempli nec nostri saeculi est.

NOTES

PLINY. *Letters*

I. (Book III, 14)

The heading of Pliny's letters is always the same (except that those to Trajan are merely *C. Plinius Traiano Imperatori*, 'Pliny to the Emperor Trajan'). *S.* stands for *salutem dat*, i.e. 'Pliny sends greetings (lit. health) to his (friend) Acilius'.

1. **Rem...dignam**] 'monstrous treatment, deserving to be described in more than a letter', lit. and not worthy of a letter only.

qui...meminisset] 'one who remembered too little, nay rather, too well, that his father had been a slave'. Instead of giving him a fellow-feeling with them, it made him more cruel still. The subjunctive is 'generic' or consecutive: 'the kind of man who'.

servisse = *serviisse.*

2. **Lavabatur**] 'he was bathing'.

exanimem] Supply *eum esse.*

an = *num.* Note the change of tenses in this sentence; the historic present, translated as a past tense, often takes historic sequence.

fidem...inplevit] 'made them believe that his death had been accomplished', lit. fulfilled the belief of the accomplished death. With *non sentire simulabat* supply *se.*

3. **quasi...effertur**] 'he was carried out as though he had fainted from the heat'.

et...et] 'both...and'.

vivere se confitetur] 'he showed that he was alive'.

4. **paucis diebus...**] 'he was with difficulty kept alive (lit. revived) and died within a few days...being avenged during his lifetime (*vivus*) as (only) the dead usually are'. The slaves would be put to death under torture.

5. **nec est, quod...**] 'and there is no reason why (*quod* is adverbial accusative) anyone can be free from anxiety, because he is...'. *possit* is an indirect 'deliberative' subjunctive (e.g. 'what am I to do?') and *sit* is subjunctive as being what the subject of *possit* thinks, i.e. 'virtual' Oratio Obliqua. See Martial, VIII, 5.

non enim iudicio...] 'for masters are murdered not after reflection (on the part of their slaves) but in wanton crime'.

Verum haec hactenus] 'but so much for this', lit. (let) this (go) so far.

6. Quid novi (est)?] 'what news is there?' partitive genitive, lit. what of new.

alioqui subiungerem] 'otherwise I should be adding it', i.e. if there were any: *alioqui* takes the place of *si quid esset.*

charta superest] 'there is still room on the page', lit. paper is left over; the paper was a roll of papyrus.

Addam, quod...] 'I will add (an incident) which appropriately occurs to me'.

in publico] 'in the public baths'.

ut exitus docuit] 'as the result proved'.

7. ut transitum...admonitus] 'being lightly touched (lit. warned)...with his hand that he might make way for him'.

quasi per gradus quosdam] 'as though by stages'.

locus] 'the scene'.

II. (III, 16)

1. Adnotasse videor...] 'I think that I have remarked that (of) the deeds and words...some are more famous, others are more truly great (*maiora*)'. *facta dictaque* are divided into *alia...alia* and are not put in the genitive case.

hesterno Fanniae sermone] 'by my conversation with Fannia yesterday'. The people concerned in this letter are as follows:

Caecina Paetus m. the elder Arria

a son (unnamed)

the younger Arria m. P. Thrasea Paetus

Fannia

2. Neptis haec (est)...] 'she (*haec*) is the granddaughter of the famous (*illius*) Arria who both consoled her husband in death and also showed him the way'. *solacium* and *exemplum* could also have been predicative dative. When her husband was involved in the conspiracy of Scribonianus, governor of Dalmatia, against Claudius in A.D. 42, Arria accompanied him to Rome and on his condemnation stabbed herself, then withdrew the dagger from her breast and handed it to him with the words, *Paete, non dolet,* 'it doesn't hurt, Paetus'. See Martial, IV.

Multa referebat...] 'she related many stories about her grandmother not less great than this one, but less well known, which I think will cause as much admiration (lit. will be as marvellous) to you when you read them as they did to me when I heard them'.

3. **eximia pulchritudine (puer...)]** Ablative of quality: 'a boy of remarkable beauty and equal modesty, and dear (lit. not less dear) to his parents as much for other reasons as because he was their son'.

4. **Huic illa ita...]** 'she...arranged the funeral procession for him without her husband's knowledge', lit. in such a way that he did not know.

quotiens intraret would be indicative in classical Latin prose because it expresses repeated action.

interroganti (ei)...] 'when he asked (lit. to him asking) how the boy was, she replied...'. *quid agit* means 'how is he?'

libenter] 'with a good appetite'.

5. **cum diu cohibitae...]** 'when the long restrained tears overcame (her)...she used to go out'.

se dolori dabat...] 'she indulged her grief; when she had satisfied it, she used to return...with a composed face, as though she had left her bereavement outside the door'.

6. **Praeclarum quidem...]** 'that indeed (was) a glorious deed of hers, to draw...'.

vocem] 'a saying'.

ista (ei) facienti, dicenti...] The datives depend on *oculos*: 'before her eyes as she did and said that was (the vision of) undying fame'. *gloria et aeternitas* is 'hendiadys' for *aeterna gloria.* 'Hendiadys' is the use of two nouns instead of an adjective and a noun: e.g. 'bread and butter' is the same as 'buttered bread'.

quo maius est...] 'it is a greater deed than this...to continue (*adhuc*) to play the mother (*matrem agere*) without the reward of immortality...'.

7. **arma moverat]** 'had taken up arms'.

in partibus] 'on his side', in the rebellion.

trahebatur] 'was being taken to Rome as a prisoner'.

8. **ut simul inponeretur]** 'that she should be taken on board with him'.

Nempe daturi estis...] 'surely you intend to give an ex-consul a few slaves (lit. young slaves) from whose hands he may receive food...'; a final relative clause.

omnia sola praestabo] 'I alone will perform all (these tasks)'.

9. **minimo (= *in minimo navigio*):** 'in a tiny vessel'.

Eadem apud Claudium...] 'she also said...in the presence of Claudius'.

profiteretur indicium] 'was turning king's evidence', i.e. giving evidence against the conspirators. This woman's name was Vibia.

audiam is deliberative subjunctive, here used in an indignant question: 'Am I to listen to you, in whose arms..., and yet (*et*) you continue to live?'

Ex quo manifestum est...] 'from this it is clear that her decision to meet a most honourable death was not suddenly made by her'.

10. deprecaretur, ne mori pergeret] 'was begging her not to continue in her resolve to die'.

si mihi pereundum fuerit] 'if I have to die', future perfect, like *vixerit* ('if she has lived') below; *mihi* is dative of agent with a gerund. He was condemned to death for treason in A.D. 66, 24 years later, but persuaded his wife, the younger Arria, to live on for her daughter Fannia's sake, instead of dying with him.

quam (=*quanta*): 'in as much harmony as I have lived with Paetus, then I do wish it'.

11. curam suorum] 'the anxiety of her family'.

sensit et...] 'she realized (this) and said: "You are wasting time (lit. doing nothing), for you can cause me to die painfully (*male*), but you cannot prevent me from dying"', lit. cause it that I fail-to-die. *ne* is not used in this final clause, as *non moriar* is taken as one word.

12. cathedra] 'from her chair', ablative of separation.

adverso parieti...] 'struck her head with great force straight against the wall'.

quamlibet duram...] 'some way of dying, however hard, if you refused me an easy one'. *negassetis*=*negavissetis*, representing in Oratio Obliqua the future perfect of direct speech.

13. Videnturne (=*nonne*)...] 'don't you think that these actions are more truly great than the famous "It doesn't hurt, Paetus", to which she came (impersonal passive) through them?' The sentence *Paete, non dolet* is used as a noun in the ablative case. Possibly her family relaxed their attempts to stop her suicide when they saw her determination to die with her husband.

cum interim...] 'while all the time immense renown spreads the one deed abroad, but no renown spreads these other deeds abroad'.

Unde colligitur...] 'from which it can be (lit. is) inferred, as I said at the beginning, that some actions...'.

III. (III, 21)

1. moleste fero] 'I am sorry'. For a life of the poet, see the Introduction to Martial (pp. 1–4).

acer] 'of keen intellect'.

qui...haberet] 'one who had in his writings both very much wit and satire and also no less sincerity'. The subjunctive is 'generic' or consecutive: 'the kind of man who...', and the genitives are partitive.

2. Prosecutus eram...secedentem] 'I had paid him the compliment of giving him a parting gift (or it may mean 'money for his journey') on his departure', for Spain, in about A.D. 98.

dederam hoc amicitiae] 'I had given this (as a tribute) to our friendship'.

3. Fuit moris antiqui] 'it was the custom in olden days', lit. it was (part) of the old custom: a partitive genitive.

honoribus] 'civil honours'.

nostris temporibus...] 'in our times, like other honourable and excellent customs, this one in particular has died out'.

laudanda] 'praiseworthy deeds'.

laudari...putamus] 'we think it is unsuitable for praise to be given at all'.

4. Remitterem...nisi (memoria) tenerem] 'I would refer you to the book itself, if I did not remember some of them', which I will quote in this letter.

si placuerint hi] 'if these please you', lit. shall have pleased.

5. mandat (ei) ut...] 'he bids her'.

Esquiliis] Local ablative: 'on the Esquiline'. The epigram quoted is Martial x, 19, 12–end. For the metre (hendecasyllables) see p. xiii.

ne...videto] 'take care that you do not knock at his learned door at an unsuitable (*non tuo*) time, when you are drunk'.

tetricae Minervae] i.e. to serious study, as Minerva was the goddess of learning.

dum centum...] 'while he prepares for the ears of the centumviral court (lit. of the hundred men, where cases of wills and property were heard: Pliny practised there) speeches (*hoc*) which future generations and posterity can ('generic' subjunctive) compare even to the pages of Arpinum', i.e. of Cicero, who was born at Arpinum.

Seras ad lucernas] 'at the time of the late-burning lamps', after dinner.

madent] i.e. with unguents.

Tunc me vel...] 'then let even (*vel*) men like stern Cato read me'. Cato the Censor, the strict moralist (*c.* 200 B.C.), is meant. Note the plural for 'men like Cato'.

6. **Meritone...dimisi...**] 'did I not (=*nonne*) rightly say good-bye on that occasion in a very friendly way to a man who ...and now that he is dead mourn for him as my dearest friend?'

quantum...potuit] 'as much as he could'.

daturus amplius...] 'and he would have given more if he could'. *daturus = et dedisset.*

aeternitas harum] 'immortality for them', i.e. for glory and praise.

(ea) quae scripsit] 'his writings'. Pliny was of course wrong in thinking that Martial's poems would not survive.

non erunt fortasse...] 'perhaps they will not, but he wrote them thinking that (lit. as if) they would be immortal'.

IV. (VI, 10)

1. **socrus meae**] Pompeia Celerina, the mother of Pliny's second wife, with whom he remained very friendly after her daughter's death in A.D. 97.

quae Rufi Vergini fuit] 'which once belonged to Verginius Rufus'. This truly great man was born, like Pliny, at or near Comum, and was Pliny's guardian. He twice refused the imperial throne offered him by his soldiers, once after he crushed the revolt of Vindex at Vesontio in 68, and again in 69, when Otho, whom he supported, was defeated by Vitellius. He was three times consul, and spent his last years in retirement, until his death, caused by a fall, in 97 at the age of 83, which Pliny describes in *Ep.* II, 1.

optimi...desiderium] 'my yearning for that best and greatest of men': objective genitive.

2. **Quocunque me contulissem**] 'wherever I went'; this indefinite clause would be indicative in classical Latin prose.

illum animus...requirebant] 'my mind, my eyes kept looking for him'.

Libuit (mihi)...et paenituit (me)] 'I wanted to see...and yet (*et*) I was sorry that I had seen it'.

3. **nec difficultas...**] 'and the difficulty of the work, which is small or (lit. and) rather tiny, is not to blame (*in causa*), but the neglect of the man to whom the task was assigned', i.e. Verginius' heir, who sold the estate and neglected the tomb. Verginius left no sons.

Subit indignatio...] 'indignation and pity comes over me (at the thought) that ten years after his death (lit. after the tenth year of his death) the remains and ashes (of a man) whose memory has spread throughout the world with glory should lie neglected...'.

pervagetur is subjunctive in Oratio Obliqua, and also because it is 'generic', i.e. 'the kind of man whose...'.

4. mandaverat caveratque] 'had given instructions and had provided', perhaps in his will. The 'deed' referred to is his refusal of the imperial throne in 68, when he defeated the rising of Vindex. The epitaph is an elegiac couplet.

Hic situs est] 'here lies', lit. is placed.

imperium...patriae] 'claimed the empire not for himself, but for his country'.

5. tam parata oblivio...] 'so soon (comes) forgetfulness of the dead', lit. so ready is it.

ut debeamus] 'that we have to...'.

cui non est verendum...] 'who cannot help fearing (lit. must not fear) what we have seen happen to Verginius?' *cui* is dative of agent with a gerundive.

cuius...facit] 'in his case his own (*ipsius*) great fame makes the wrong more conspicuous in proportion as it is more disgraceful', lit. as more disgraceful, so also more conspicuous.

V. (VI, 16)

1. Petis (a me)] 'you ask me'.

avunculi] This uncle was Gaius Secundus Plinius, born (like his nephew) at Comum in A.D. 23, later becoming the adoptive father of the younger Pliny who, according to custom, took his name. He served both in the army and in civil duties as an *eques*, and spent all his spare time in writing on a variety of subjects, the only books which survive being his great *Natural History*, dedicated to the emperor Titus, whose father Vespasian was his great friend. *scribam* is 'write an account of'.

quo verius...possis] A final relative clause: 'so that you may be able to give a more accurate account to posterity', i.e. in his *Histories*, but the part containing a description of the eruption has not survived.

Gratias ago] Supply *tibi*.

video morti eius...] 'I see that immortal glory is bestowed upon (or 'has been promised to') his death, if it is celebrated...'. *celebretur* is subjunctive in Oratio Obliqua. *morti* *inmortalem* is an instance of 'oxymoron', the combination of opposites to make a striking effect.

2. **quamvis...occiderit]** 'although he perished in a death that will as it were make his name live for ever (*victurus* is from *vivo*; lit. as one destined to live for ever: another 'oxymoron'), in a disaster involving (lit. of) a most beautiful district, in the famous destruction both of a people and of a city', i.e. Pompeii. The text here is not quite certain. *quamvis* in Ciceronian Latin generally means 'however much'.

quamvis...addet] 'although he himself composed many works which will survive (*et* is omitted in English; in fact only his *Natural History* survives), yet the immortality of your writings will add much to his everlasting fame'.

3. **beatos (esse eos)...]** 'that those men are happy to whom it has been granted...to do deeds worth writing about or to write books worth reading, yet most happy are those to whom both favours (have been granted)'. Supply *datum est* with *utrumque*.

Quo...] 'wherefore I undertake more willingly, in fact (*etiam*) I demand, the task which you lay upon me'.

4. **Miseni]** Locative.

classemque...regebat] 'and he was commanding the fleet in person with supreme authority'. One of the two fleets defending Italy was stationed at Misenum, the other at Ravenna. The commander was an imperial *praefectus* of equestrian rank, whose duties kept him often at Rome.

Nonum Kal. Septembres=*ante diem nonum*: i.e. 24 August A.D. 79. The day between sunrise and sunset was divided into 12 hours of equal length, varying with the season: the seventh hour in August would be about 1 p.m.

mater mea] The admiral's sister, Plinia.

inusitata specie] Ablative of quality: 'of unusual appearance'.

5. **Usus ille...]** 'he had been sun-bathing, then bathing in cold (water, supply *aqua*), and had lunched while lying down'. Romans in the house often went bare-footed, as Pliny did.

maxime] 'best'.

incertum (erat)...] 'it was uncertain to those looking at it from a distance, from which mountain, but it was afterwards found that it was Vesuvius'. The volcano is 18 miles from Misenum, with other hills both between them and beyond.

non alia...] 'no tree could represent better than the pine'. *expresserit* is 'potential' subjunctive.

6. longissimo velut trunco...] 'rising into the sky (*altum*) with what may be called (*velut*) a very long trunk it spread out into several (lit. certain) branches, because, I suppose, it was lifted by a recent blast of air; then when the blast died down (ablative absolute) being left to itself or even overcome by its own weight, it broadened out before vanishing (lit. vanished in the direction of breadth)...'.

7. Magnum...visum (est ei)] 'it seemed to him, as being a man of great learning, to be an important phenomenon and one to be examined closer at hand'.

mihi...facit copiam] 'he gave me the chance (of coming), if I wanted to come too (*una*, an adverb)'. *vellem* is subjunctive in a conditional clause in Oratio Obliqua, depending on the historic present *facit*.

quod scriberem dederat] 'he had given me some writing to do', lit. (something) which I might write; a final subjunctive.

8. Egrediebatur...accipit...] We should say 'when he was leaving...he received a note from Rectina, (the wife) of Tascus, who was terrified by the impending danger'.

subiacebat] i.e. at the foot of Vesuvius.

nec ulla (erat)...fuga] 'and there was no means of escape except by ship'.

discrimini] 'from so great a danger': dative after a verb of taking away.

9. Vertit...maximo] 'he changed his plans, and carried out in a heroic (*maximo*) spirit what he had undertaken in a scientific spirit'. It was now a very risky undertaking to approach the volcano.

Deducit quadriremes] 'he had the quadriremes launched'. They were beached when not in use.

erat frequens amoenitas orae] 'the pleasant shore (=*amoena ora*) was thickly populated'.

laturus auxilium] 'to bring help not only to Rectina...'.

10. rectum...tenet] 'held a straight course, a straight helm, directly towards the danger'.

omnes...oculis] 'all the changes, all the phases of that disastrous eruption (*mali*, a noun) as he noticed them with his eyes'.

dictaret enotaretque] 'dictated and gave notes on...', to a secretary. The elder Pliny made a fetish of taking notes, and left his nephew 160 closely written volumes.

11. quo...densior] 'hotter and thicker, the nearer they came'. In classical prose *accederent* would be indicative, and *eo* would precede *calidior*. The verb for *pumices* and *lapides* is *incidebant*, for *vadum* and *litora* supply *erant*: 'the water suddenly became shallow, and the shores were blocked (lit. blocking) by fragments (*ruina*) of the mountain'.

Cunctatus an (=*num*) 'After hesitating whether to turn back, he said to the helmsman who advised him to do so: "'Fortune favours the brave', make for Pomponianus"', one of his friends.

12. **Stabilis erat...**] 'he (i.e. Pomponianus) was at Stabiae (4 miles south of Pompeii), separated by the intervening (*medio*) bay'.

nam sensim...] 'for the sea flows into a bay which gradually sweeps round in a curve', lit. shores which are sent round and are curved.

quamquam is here used with the ablative absolute instead of the indicative: 'though the danger was not yet approaching, yet as it was obvious and, since it was increasing, was becoming pressing (*proximo*)...'.

certus fugae] 'determined on flight (objective genitive; a poetical construction) if the contrary wind should drop'. *si resedisset* is in 'virtual' Oratio Obliqua depending on *certus fugae*. His own words were 'I shall flee if the wind drops': future perfect.

quo (vento) secundissimo invectus] 'sailing with the wind right behind him'.

trepidantem (eum)] 'embraced Pomponianus, who was in a state of panic'.

deferri (se) iubet] 'gave orders that he should be taken to the baths'.

lotus] 'after bathing'.

aut...similis hilari] 'either being quite cheerful or, what is equally praiseworthy (*magnum*), pretending to be cheerful'.

13. **pluribus locis**] Local ablative without *in*: 'in several places'.

excitabatur] 'was set off' or 'increased'.

in remedium formidinis] 'to calm his friend's fear he kept saying that fires were burning that had been left in the panic of the country-folk, and houses were on fire that had been deserted in the abandoned districts (*per solitudinem*)'.

verissimo somno] 'with a perfectly genuine sleep'.

meatus animae] 'his breathing, which was somewhat (comparative) heavy and loud...'.

14. **ita...surrexerat**] 'had become so choked up with...(lit. being filled up had risen) that if he remained any longer... (supply *esset*: it is subjunctive because the conditional clause

depends on a consecutive clause) the exit would be prevented (lit. denied)'.

15. **se reddit...**] 'rejoined Pomponianus'.

In commune] 'together'. Supply *utrum* before *intra*: 'whether they should...'.

intra tecta] 'inside the building'.

quasi emota...videbantur] 'the house seemed to move backwards and forwards as though it had been shifted from its foundations'.

16. **quamquam** goes only with *levium*: 'in the open air on the other hand (*rursus*) there was the danger of the falling (lit. the fall was feared) of pieces of pumice stone, light and fine (lit. eaten away) though they were, but after a comparison of the risks they chose this latter course', lit. which a comparison...chose.

apud illum quidem...] 'in his case indeed one calculation outweighed the other', i.e. the admiral really chose the wiser course, the others rushed out of doors in a panic.

Cervicalia...] 'they placed pillows on their heads and fastened them with napkins; that was their protection against falling objects'. *id* is 'attracted' to agree with *munimentum*.

17. **Iam dies (erat)...**] 'by this time it was day elsewhere, but there it was night darker and deeper than any (lit. all) night'.

solabantur] 'relieved', i.e. lighted up.

Placuit (ei)...] 'he determined to go out', lit. it pleased him.

ex proximo...] 'to see from close at hand whether (=*numquid*) the sea allowed any (ships to put out) yet'.

quod...permanebat] 'but it still remained rough and contrary'.

18. **abiectum linteum**] 'an abandoned piece of cloth', perhaps a sail: larger of course than the *lintea* of § 16.

semel atque iterum] 'once or twice'.

frigidam] Supply *aquam*.

flammarum praenuntius] 'which heralded flames'.

19. **servulis**] The diminutive probably means 'young slaves', as in II, 8.

concidit, ut ego colligo...] 'he fell dead, because, I imagine (lit. as I infer), his breathing was hindered and his windpipe (*stomacho*) choked by the thickness of the vapour (lit. by the too thick vapour); his wind-pipe (lit. which to him; the antecedent is *stomacho*) was naturally weak and obstructed (*angustus*) and often inflamed'.

20. **Ubi dies redditus (est)**] 'when the light of day returned

—that was the third day from the one when he had last (*novissime*) seen daylight—his body was found whole and uninjured and covered over, fully dressed (lit. just as he had been dressed): the appearance of his body was more like that of one sleeping than one dead'.

tertius (dies)] This is the Roman 'inclusive' method of counting, whereby the first and last of a series is counted; we should call it 'the second day'. The eruption was on 24 August, there was no daylight on the 25th, and he was found on the 26th.

21. ego et mater] Supply *eramus*.

nihil ad historiam] Supply *hoc pertinet*: 'this does not concern your history'. *exitu eius* is of course 'my uncle's death'.

22. Unum adiciam...] 'I will only add that I have related all the events which I witnessed myself (lit. at which I was present), and which I heard about at once, when the truth is generally (*maxime*) told'.

potissima excerpes] 'you will select the most important facts'.

aliud...aliud] 'for it is one thing to write a letter, another a history, one thing to write to a friend, another for the general public'.

VI. (VI, 20)

1. Ais te...] 'you say that you have been interested by the letter, which I wrote to you at your request (*exigenti*)...and wish to know not only what fears but also what dangers (*casus*) I endured when left behind at Misenum, for after reaching that point I broke off my story'.

quamquam...incipiam] Aeneas' words to Dido in Vergil, *Aeneid*, II, 12: 'although my spirit shrinks from the memory...', lit. trembles to remember.

2. mox balineum] Supply *sequebatur*: 'then followed...'.

3. minus formidolosus, quia Campaniae solitus (est)] 'causing less terror (than it would elsewhere) because it is usual for Campania'. In A.D. 63 much of Herculaneum was destroyed by an earthquake.

ita invaluit...] 'it became so strong that everything seemed (lit. was believed) not (only) to be moved but to be overthrown'.

4. **surgebam...]** 'I was just getting up myself (*in vicem*), intending to rouse her if she was asleep'. *quiesceret* is subjunctive in 'virtual' Oratio Obliqua depending on *excitaturus*.

area] This courtyard was not surrounded on all sides by buildings, but was open to the sea on one side.

5. **Dubito...]** Supply *utrum*: 'I am not sure whether I ought to call (my behaviour) courage or folly'.

agebam enim] Supply some words like '(it was probably folly) for I was only in (*agebam*) my eighteenth year'. Hence he was born in A.D. 61 or 62.

Titi Livi is Livy the historian.

quasi per otium] 'I read it as though quite at ease, and even made some extracts, as I had just begun to do', thus following his uncle's example.

Ecce] Translate 'suddenly'. Pliny's uncle had been procurator of Spain.

ut] 'when'.

patientiam...corripit] 'upbraided her for her indulgence (i.e. in allowing me to keep her there) and me for my carelessness of danger'.

Nihilo segnius] Supply *manebam*: 'I remained intent on my book as studiously as before', lit. by nothing more slackly.

6. **hora prima (fuit)]** About 6 a.m.

adhuc dubius...] 'the daylight was still doubtful and somewhat (*quasi*) feeble'.

circumiacentibus is an adjective: 'when the surrounding buildings were shaken, there was a great and certain danger (lit. fear) from the falling masonry (*ruinae*), (as we were) in a place which though open was small', i.e. the courtyard of the house.

7. **visum (est nobis)]** 'we determined', lit. it seemed (best).

sequitur vulgus...] 'a terrified crowd followed us, and, (taking a course) which in panic seems to be the wise one (lit. which (is) like to prudence), preferred somebody else's plan to its own, and pressed upon us in a huge body as we departed (*abeuntes*: supply *nos*) and urged us forward'.

Egressi...consistimus] 'after making our way outside the buildings (of the town) we stopped'. Note that *egredior* here takes the accusative instead of *ex* with the ablative.

8. **Multa miranda, multas formidines]** 'many marvellous things, many perils'.

quamquam is again used without a verb, as also in § 6 above.

in contrarias partes...] 'kept moving in different directions, and not even when stopped (*fulta*) by stones did they remain stationary', lit. in the same track.

9. **in se resorberi**] 'sucked back'.

Certe processerat...] 'at any rate the shore had advanced (i.e. by the withdrawal of the sea), and was keeping back many sea-creatures on the dry sands'. *harenis* is local ablative.

Ab altero latere...] 'on the other side (i.e. the landward side) a black and dreadful cloud, which was riven by twisted and quivering flashes (*discursibus*) of fiery vapour, was splitting into long flaming shapes', lit. long shapes of flames.

similes et maiores] 'similar to and larger than lightning flashes'.

10. **acrius et instantius**] 'more firmly and insistently'.

frater, avunculus] He addresses Pliny's mother and Pliny himself in turn.

superstites voluit (vos esse)] 'he wanted you to survive him'.

quid cessatis] *quid* often equals *cur*: 'why do you hesitate to escape?'

non commissuros nos, ut...] 'that we would not allow ourselves to consider our own safety (supply *saluti*) while we were still uncertain about his safety'. *committere ut* means 'to act in such a way that...'.

11. **ultra**] 'any longer'.

proripit se...] 'he rushed off and hurried away (lit. was carried away) from danger at full speed'.

Nec multo post...] 'not long afterwards that cloud (which I mentioned before, *illa*)...'. *descendere, operire* are historic infinitives, which usually describe a succession of events.

Miseni quod procurrit, abstulerat] 'had hidden from our sight the promontory of Misenum', lit. what of Misenum runs out (into the sea): partitive genitive.

12. **orare, hortari, iubere**] Historic infinitives; supply *me*. They are followed by the jussive subjunctive, *quoquo modo fugerem*, 'that I should escape by any (lit. whatever) means'.

(me) posse iuvenem...] This is what she said: 'I could do so (i.e. escape) being young, but she, being hindered (*gravem*) both by her years and her physical disabilities (*corpore*: she was apparently corpulent like her brother), would die happily, if she was not...'.

Ego contra...] 'I on the other hand said (supply *dixi*) that I would not escape except with her'. *una* is an adverb.

addere gradum] Supply *gradui*: 'seizing her hand I forced her to step forward', lit. add step to step.

aegre] 'unwillingly'.

quod me moretur] 'for delaying me'. The subjunctive is used because this was the reason which she gave.

13. cinis] Supply *incidebat*: 'by this time ash was falling, but still a fine ash'.

torrentis modo] 'like a torrent', lit. in the manner of.

infusa terrae] 'pouring over the ground'; *terrae* is dative governed by *infusa*, which is used in the passive like a deponent verb.

Deflectamus...] Supply *via*: 'let us turn aside from the road, while we can see, so that we may not be overthrown (*strati*) and trampled on by the crowd of those who are accompanying us'.

14. Vix consederamus, et nox (=*cum nox venit*). Pliny often uses *et* where we use 'when'.

non qualis inlunis...] 'not as it is when there is no moon... but as it usually is in a closed room (*locis*) when the light is put out'. *est* is twice omitted in this clause.

Audires] 'one could hear': potential subjunctive, the second person singular being used 'generally', as often in English.

vocibus requirebant...] 'were calling out for their parents (lit. were looking for them with cries)...or recognizing them by their voices'. *vocibus* is used in two different meanings.

hi suum...] 'some were lamenting their own fate, others that of their families'.

erant qui...] 'there were some who...': a 'generic' or consecutive subjunctive.

15. tollere is historic infinitive, which is rarely used unless there are more than one in the sentence.

plures...interpretabantur] 'the majority thought that there were (supply *esse*) no longer any gods anywhere and that this was the last (*novissimam*) and eternal night for the world'.

Nec defuerunt qui...] 'and there were not lacking some who...': another 'generic' *qui*, but in *aderant, qui...nuntiabant*, the indicative is used as referring to certain definite people; 'there were present some who declared falsely, but finding credulous listeners (lit. to people believing it), that at Misenum one building (*illud*) had collapsed, and another (*illud*) was on fire'.

16. Paulum reluxit...] 'a little light appeared again; but it (*quod*) seemed to us to be not daylight but the sign of approaching fire'. *reluxit* is impersonal.

longius substitit] 'stopped some way off'; note the comparative.

tenebrae] Supply *venerunt*.

operti essemus] 'we should have been covered'.

17. Possem...nisi...credidissem] 'I might have boasted that no groan, no cowardly words (*vocem parum* (=*non*) *fortem*) fell from my lips, had I not believed that I was perishing with the whole world, and that the whole world (*omnia*) was perishing with me, which brought me a melancholy but great comfort in my approaching death', lit. with miserable but great comfort of death.

18. **dies verus (venit)**] 'genuine daylight came'.

cum deficit] 'when there is an eclipse', lit. when it (i.e. the sun) fails.

Occursabant...] 'everything met our eyes, which were still looking round in terror, changed and covered with ash as though with snow'.

19. **curatis utcunque corporibus**] 'after looking after our bodily needs as best we could', i.e. after bathing, eating, etc.

suspensam dubiamque noctem] 'a night of suspense and doubt'.

plerique lymphati] 'many people in their madness'. *vaticinationibus* goes with *ludificabantur.*

et sua et aliena mala] 'the misfortunes both of themselves and of others'.

20. **Nobis...consilium (erat)...**] 'not even then did we (i.e. Pliny and his mother) think of departing (lit. was there a plan to us), although we had both experienced and were expecting danger, until (there came) news of my uncle'.

Haec...videbuntur] 'you will read this account, but you will not use it for your book (*non scripturus*) as it is by no means worthy of (a place in) history, and you will blame yourself (lit. put it down against your account, like a bill) if you do not think it worthy even of a letter, because of course (*scilicet*) you yourself demanded it'.

VII. (VII, 27)

1. **Et mihi discendi...**] 'leisure gives an opportunity both to me of learning and to you of imparting information'.

perquam velim scire] 'I should very much like to know': potential subjunctive. Supply *utrum* after *scire*: 'whether you think that ghosts really exist (*esse*) and have a form of their own and something supernatural, or (being) imaginary and false take on an appearance (*imaginem*)...'.

2. **ut esse credam...**] 'I am particularly induced to believe that they do exist by the experience (*eo*) which I hear befell Curtius Rufus'. All our knowledge about this man comes from Pliny and from Tacitus, who tells the same story in *Annals*, XI, 21.

Tenuis adhuc...] 'while still poor and unknown he had accompanied as a member of his staff (*comes*) the man who was governor of Africa'. Supply *viro* with *oblinenti*.

Inclinato die] 'when the sun had just passed midday'.

offertur ei...] 'there appeared before him the form of a woman of superhuman size and beauty', lit. larger than human.

perterrito (ei)...] 'she said to him in his terror that she was (the genius of) Africa and was foretelling the future'; the next sentence is still what she said: 'for he (*eum* understood) would ...hold high office (*honores*)...'.

cum summo imperio] 'with supreme command'.

3. Facta sunt] 'came true'.

accedenti agrees with *ei* understood and *nave* is governed by *egredienti*, 'as he was approaching Carthage and disembarking ...the same figure...met him'.

futura...auguratus] 'inferring future events from the past, adversity from prosperity'.

nullo...desperante] Ablative absolute: 'though none of his family despaired'. *salutis* is 'recovery'.

4. illud] 'the following story'.

ut accepi] 'just as I heard it'.

5. Erat Athenis] 'there was at Athens'. *ferri* is a noun.

si attenderes acrius] 'if one (lit. you) listened more carefully'. When 'you' is indefinite the subjunctive can be used in such conditional clauses.

longius...reddebatur] 'proceeded at first from a distance, then from close at hand'.

promissa barba] 'with a long beard': ablative of quality.

gerebat] 'he wore'.

6. inhabitantibus] Probably dative of the person concerned: 'miserable and terrible nights were spent in wakefulness by the inmates of the house'.

longior...erat] 'the fear remained longer (= *diuturnior*) than the reason for the fear'.

Deserta...] Supply *est*: 'was deserted and condemned to solitude (ablative of the penalty imposed) and left entirely (*tota*) to that horror'.

proscribebatur...] 'it was advertised, however, in case (*seu*) anyone might want to buy it or rent it in ignorance (lit. being ignorant) of such a danger'. *mali* is a noun and is objective genitive. *vellet* is in 'virtual' Oratio Obliqua.

7. suspecta] Supply *est ab eo*.

omnia docetur] 'was told everything'. The second accusative is retained after the passive with some verbs.

nihilo...magis] 'none the less, in fact all the more so'. The ablatives are 'measure of difference'.

sterni] Supply *lectum*: 'ordered a couch to be spread (i.e. with cushions) for him', so that he could sit and write.

in prima parte] 'in the front part of the house', i.e. in the *atrium*, or hall, near the front door.

pugillares were wax-coated tablets for writing notes with a metal *stilus* or pen. See illustration on page 61.

suos omnes] 'all his household'.

in interiora] 'into the inner part of the house'.

animum...intendit] 'directed his attention...to writing'. The historic present is here, as often, followed by an imperfect subjunctive.

ne vacua mens...] 'so that his mind might not be empty and so imagine phantoms, which it (thought it) had heard, and idle fears'.

8. **Initio**] *erat* is twice omitted here. 'At first there was (only) the silence of night, as there was everywhere.' Historic infinitives follow down to *audiri*.

offirmare animum...] Supply *animum* with *praetendere* also, as the direct object: 'he firmly fixed his attention and thus closed his ears', lit. interposed his attention before his ears.

ut] 'as though'.

narratam sibi effigiem] 'the figure which had been described to him'.

9. **similis vocanti**] 'as though calling him', lit. like to one calling.

Hic contra...] 'Athenodorus, however, made a sign with his hand that the ghost should wait a little'.

scribentis (eius) capiti] 'it rattled its chains over his head as he continued to write'.

idem quod prius innuentem] 'looked back at the ghost which was making the same gesture as before'.

nec moratus...] 'without delay he picked up the lamp...'.

10. **gravis vinculis**] 'weighed down with chains'.

deflexit] Supply *se*.

Desertus] 'being thus left alone he put down some grass...to mark the place'. *signum* is of course in apposition to *herbas* and *folia*.

11. **magistratus** is accusative plural, and *eos* understood is the object of *monet*.

quae corpus...] 'which the body, decomposed by its long time in the earth (lit. by time and earth), had left bare and eaten into by the chains'.

collecta...] 'they were gathered up and buried at the public expense'.

12. **Domus postea...**] 'the house was then free from the ghost (*mānibus*), which had been duly laid to rest'. Burial with due rites was considered necessary for the repose of the soul.

Et haec quidem...] 'I indeed believe those who tell this story, and I can tell the following story in my turn', lit. to others.

Est libertus mihi...] 'I have a freedman who has received some education', and whose story was therefore likely to be true.

minor (natu)] 'younger'. Note the omission of *in* with *eodem lecto*.

Is visus est sibi cernere] 'the younger brother thought that he saw...', lit. seemed to himself to see.

circa verticem tonsus reperiuntur...] 'he was found with the hair on his head cut short, and the hair was found lying (on the floor)'.

13. **Exiguum temporis medium (erat) et...**] 'a short space of time (partitive genitive) intervened when (lit. was between, and) another similar event gave confirmation to the first one'.

mixtus pluribus] 'together with several others'. This was of course another boy, not the one whose hair had been cut off the first time.

in tunicis albis duo...] 'two men clad in white tunics...and cut his hair as he lay there (supply *eum* with *cubantem*) and retired by the way by which they had come'.

Hunc...ostendit] 'daylight showed that he too...and revealed his hair scattered about'.

14. **Nihil...secutum (est)...**] 'nothing remarkable followed except perhaps that I was not accused (*reus*), though I would have been (*futurus = sed fuissem*) if...'.

datus...libellus] 'an information about me given him by Carus'. Mettius Carus was a well-known informer.

ex quo...] 'from this it may be inferred that because it is the custom (*moris est*, partitive genitive) for accused men to let their hair grow long, the fact that my slaves' hair was cut short was a sign of my escape from the danger which threatened me'. Note the two instances of a noun and a past participle passive used where we use two nouns or a clause. Similarly *ab urbe condita* is 'from the foundation of the city'. Pliny's explanation of this strange phenomenon is not very convincing and may have been merely an excuse for mentioning his escape.

15. **rogo...intendas**] 'I ask you to apply your learning (to the question)': jussive subjunctive.

Digna res...] 'the subject deserves your long and careful consideration (lit. is worthy which you should consider, a 'generic' relative clause), nor indeed am I unworthy of having your knowledge placed at my disposal', lit. not even am I unworthy to whom you should give the chance of your knowledge.

16. **Licet...disputes**] 'though you may argue on either side, as you usually do, yet argue (supply *disputa*) more strongly on one side (*ex altera*), so that you may not send me away in suspense and uncertainty, since my reason for consulting you was that I should cease to have any doubts'.

VIII. (IX, 23)

1. **Frequenter agenti mihi...**] 'it has often happened to me while pleading in court (*agenti*) that the members of the Hundred Court (which was concerned with cases of wills, property, etc.), after restraining themselves as befitted (lit. within) the position and dignity of judges, have all suddenly stood up and applauded (me), as though overcome and forced to do so'.

2. **famam...rettuli**] 'I have won as much (supply *talem*) fame as (*qualem*) I had ever (lit. most greatly) hoped for'.

cepi] 'I have received'.

ex sermone Taciti] 'from a conversation with Tacitus'. This was his friend the famous historian, to whom he addressed the letters about the eruption of Vesuvius.

secum...equitem] 'that a Roman *eques* sat next to him at the last games'. Tacitus was of senatorial rank, and so must have chosen to sit on this occasion with the *equites*, who could not use the senatorial seats. The 'games' were chariot races.

Hunc...requisisse (=*requisivisse*). This and the following sentences are what Tacitus told Pliny; 'after much learned conversation this man asked Tacitus, "are you...?"' Supply *utrum* before *Italicus*, as also before *Tacitus* below.

Nosti (=novisti): 'you know me, and from my literary work too'.

3. **Ad hoc illum (rogavisse)**] 'at this the other asked'.

quam sit iucundum...] 'what pleasure it gives me (lit. how pleasant it is) that (*quod*) our names are ascribed to literature, as though they belonged (*propria*) to literature, not to men, and that each of us is known through (*ex*) his writings even to those (*his*) to whom he is otherwise unknown'.

4. **Accidit aliud...**] 'another similar event happened a few days ago' = *pauculis ante diebus*.

Recumbebat] i.e. on the same couch at dinner. There were three couches round a table (the fourth side being left free for service), each couch accommodating three guests. On this occasion Pliny had the chief place (*locus imus*, lit. the lowest, but in fact the most honourable) on this couch, next to him was Rufinus, and then his friend.

Multa] Supply *dixit*. **Plinius est**] 'it must be Pliny'.

5. **capio...fructum**] 'I receive great recompense for my work (from this)'. *an* is merely an interrogative particle.

iure laetatus est] 'was right to be pleased'.

ita] 'by saying'.

Οὗτός ἐστι Δημοσθένης] 'this is Demosthenes'.

et...et] 'both...and'.

6. **iactantior**] 'too boastful'.

iudicium] 'the opinion'. **apud te**] 'to you'.

faves nostris (laudibus)] 'are pleased with any praise bestowed on me'. *nec...et* = 'both...not...and also'.

IX. (IX, 33)

1. **Incidi in...**] 'I came across a true story, but it is very like a made-up tale, and is worthy of that most rich, lofty, and truly poetic genius of yours'. Caninius was planning to write an epic poem on Trajan's Dacian campaigns. This same story is told by Pliny's uncle in his *Natural History*, IX, 26.

incidi autem...] 'I came across it, I say, while various miraculous events were being related by different people (*hinc inde*, lit. from here and there) at dinner'.

Magna auctori fides (est)...] 'I have great faith in my informant, and yet (*tametsi*) what has a poet to do with faith?' lit. what (is there) to a poet with belief.

Is tamen auctor...] 'yet my informant is one whom you could have fully believed, even if you had been intending to write (*scripturus* = *si scripturus fuisses*) a historical work'.

2. **Est...Hipponensis colonia**] 'there is in Africa the colony of Hippo', near Carthage and Utica.

Adiacet...] 'near it there is a lake big enough for navigation'.

in modum fluminis] 'like a river'.

vice alterna] 'alternately'.

nunc infertur...] 'at one time flows into the sea, at another returns to the lake'.

3. **Omnis hic aetas...**] 'people of all ages here are filled (*tenetur*) with eagerness for...'.

maxime pueri...] 'especially the boys, who are attracted by...', lit. whom leisure rouses.

His (est) gloria...] 'their glory and skill consists in swimming out as far as possible'.

victor (est) ille...] 'the winner is he who has left not only the shore but also (*ita...ut*) his fellow-swimmers (lit. those swimming together) the farthest behind'.

4. audentior ceteris...] 'bolder than the rest, was making his way out farther (than the others)'.

occurrit, praecedere] The indicative and historic infinitives alternate in these sections.

subire] 'took him on his back'.

trepidantem] Supply *eum*, 'frightened'.

in altum] 'into the open sea'. **reddit]** Supply *eum*.

5. tamquam miraculum] 'as though he were a marvel'.

audire, narrare] 'heard his story, repeated it'.

si quid est mari simile] 'any other water that resembled the sea', lit. if there is anything like the sea, i.e. the lake and estuary.

ille] 'this boy'.

Delphinus] Supply *venit*.

ad tempus] 'at the same time as before'.

revocet is governed by *quasi*; supply *et* after *invitet*.

mergitur] 'dived': passive used intransitively.

varios orbes...] 'indulged in various twists and turns', lit. winds and unwinds various circles.

6. Hoc] Supply *fecit*. *altero* means 'the next'.

pluribus (diebus)] 'on several days'.

donec...pudor] 'until shame at their fear seized men who were brought up from childhood in the sea'. In classical prose *donec* would here take the indicative.

praebentem] Supply *delphinum* and *se*: 'the dolphin who was offering himself (to be touched)'.

Maxime puer...] 'in particular the boy who was the first to have the experience swam up to meet the dolphin' (lit. swam up to him swimming).

fertur referturque] 'was carried backwards and forwards'.

huius] 'of the boy'. **illius]** 'of the dolphin'.

7. Nec non] 'and also'. *alii = ceteri*.

dextra laevaque] 'on right and left'.

una] An adverb: 'with him'.

tantum...] 'only as spectator...'.

Nihil simile...] 'he neither did anything nor endured anything like the first dolphin...just as the other boys were leading the boy to and fro'; supply *ducebant reducebantque*, lit. were leading him and bringing him back.

8. **Incredibile (est)...**] 'it is incredible, but as true as the previous events, that the dolphin which carried and played with the boys became accustomed to drag himself out on to the shore, and after drying himself on the sands when he grew hot used to plunge back into the sea'. The main verb of the indirect statement is *solitum* (*esse*). *harenis* is local ablative without *in*, and *incaluisset* is subjunctive of repeated actions and also because it is in Oratio Obliqua. *extrahi* and *revolvi* are passives used reflexively.

9. **Constat...**] Impersonal: 'it is a well known fact that Octavius...with a misplaced (*prava*) show of reverence poured ointment over the dolphin (supply *delphino*) when he had dragged himself up on to the shore'. He thought the fish was the incarnation of a sea-god: it was usual to anoint statues of the gods.

cuius illum...] 'the strange smell (lit. strangeness and smell, a 'hendiadys', see Pliny II, 6 note) of which he escaped (by swimming out) to sea, and was only (*nec nisi*) seen again after many days (when he was) languid and timid'. Note that this relative clause in indirect statement is in the accusative and infinitive, instead of the subjunctive, as it is equivalent to a main sentence.

solita ministeria repetisse (=repetiisse)] 'returned to his usual attentions', such as carrying the boy on his back.

10. **magistratus**] Nominative plural.

quorum adventu...] 'and by their arrival and stay (in the place) the modest resources of the town (*modica res publica*) were exhausted by new expenses', because the visiting officials were entertained at the public cost.

secretum] 'its seclusion'.

placuit...] 'it was decided (lit. it was pleasing to them) that (the creature) to which the crowds came (impersonal passive) should be killed secretly'.

11. **Haec tu...**] 'with what pathos, with what eloquence (*copia*: supply *verborum*) will you weep over this tale (*haec*), adorn it and magnify it!'

Quamquam] 'and yet'.

non est opus...] 'there is no need for you to make anything up...'.

sufficit] 'it is enough that the true facts should not be belittled'.

X. (X, 96)

1. Sollemne est...] 'it is my invariable rule to refer to you, Your Majesty (*domine*), all cases about which I am doubtful'.

regere] 'to direct'.

Cognitionibus] These were informal enquiries held by a magistrate. Note that *de Christianis* depends directly on *cognitionibus*, though in classical Latin a verb would usually be required, like *cognitiones de Christianis habitae*, 'enquiries held about Christians'.

quid...quaeri] 'what is usually either punished or investigated, and to what extent'.

2. Nec mediocriter...] 'I have felt no small doubt whether (*-ne = utrum*) there should be any difference made in respect of age (lit. of ages), or whether (people) however young (*quamlibet teneri*) should not differ at all (in their treatment) from those who are of riper years'.

detur] Supply *utrum* before this word, as also before *nomen* below.

an ei...] 'or whether it should not count in the favour of a man (*ei prosit*) who has ever (lit. at all) been a Christian that he has ceased to be one' (*desisse = desiisse*).

si careat] 'if it should be free from crimes'; the main verb, *puniantur*, is already subjunctive in an indirect question, and *careat* may therefore be subjunctive in the subordinate clause, but the indicative *fuit* is used above, so *careat* is probably a 'vague future' condition.

flagitia cohaerentia nomini] 'the crimes connected with the name', such as cannibalism and other ridiculous accusations made against Christians.

in iis...] 'in the case of those who...'. **tamquam]** 'as'.

hunc modum] 'the following method'.

3. an = *num*.

Confitentes (eos)...] 'if they admitted it I questioned them...; if they persisted I ordered them to be led off (to execution)'.

qualecunque...] 'whatever were the proceedings to which they confessed'; *dubito* in a negative sentence here takes the accusative and infinitive instead of *quin*.

4. Fuerunt alii...] 'there were others suffering from the same madness (genitive of quality) whom...I noted down (in my record of the case) to be referred to Rome'.

Mox ipso tractatu...] 'later on, in the mere course of the enquiry, as usually happens, when the accusation became widespread (lit. spread itself abroad: ablative absolute) a larger number of instances came to my notice (*inciderunt*)'.

5. **Propositus est...**] 'an anonymous (*sine auctore*) accusation (lit. little book) was put forward...'.

Qui negabant...] The antecedent of *qui* is *eos* understood, governed by *putavi*: 'I thought that those who denied...ought to be acquitted'.

praeeunte me] 'following my dictation', lit. I preceding, ablative absolute.

male dicerent Christo] 'cursed Christ'.

quorum...(ei) qui...] 'for it is said that those who are really (*re vera*) Christians cannot be compelled to do any of these things'.

6. **ab indice**] i.e. in the anonymous *libellus*.

mox] 'later'.

fuisse] Supply *dixerunt se*.

ante triennium, ante plures annos = *triennio ante, pluribus annis ante*.

non nemo...] 'some people (said they had ceased to be Christians) twenty years previously'. 'not no one' = 'some'.

7. **Hanc fuisse summam**] 'that this had been the sum total ...namely that (*quod*)...'.

stato die] 'on a fixed day', i.e. Sunday.

carmen dicere secum in vicem] 'to sing a hymn, taking parts in turn'. This refers to the early Christian practice of singing antiphonal chants.

non in scelus aliquod...] 'not to commit some crime, but to refrain from (*ne*) committing...'.

ne fidem fallerent...] 'not to break their word or refuse to return a deposit when called upon to do so'. This was a common crime when banks were few and valuables were often deposited with friends.

Quibus peractis...] 'when this ceremony was over their custom had been (*morem sibi fuisse*) to depart and meet again to take a meal (i.e. in the 'Agăpē' or love-feast), but a meal of an ordinary and innocent kind': not of human flesh, which they were accused of eating. This sentence is still indirect statement depending on *adfirmabant*.

quod ipsum...] 'they had ceased to do even this...whereby according to (*secundum*) your instructions I had forbidden any societies to exist (*esse*)'.

8. **Quo**] 'wherefore'.

ex...quaerere] 'to enquire from two maid-servants, who

were called "deaconesses", even (*et*) by torture what was the truth' (partitive genitive).

ad consulendum...] 'I hastened to consult you'.

9. **digna consultatione...]** 'seemed to merit a consultation (with you), especially because of the number of those involved', lit. of those in danger.

omnis aetatis] Genitive of quality: 'many people of every age, of every rank, of both sexes are still (*etiam*) being summoned to (answer) the dangerous charge'.

Neque tantum] 'not only'.

quae...posse] 'but I think it can be stopped and cured'.

10. **satis constat...]** 'it is pretty well agreed that the temples, which were recently almost deserted, have begun to be frequented, and the sacred rites, which were long interrupted, to be revived, and the food for victims to be sold again'. *venire* is from *veneo*.

Ex quo...] 'from this it is easy to infer what a large number... if an opportunity were to be given for repentance'. For *si sit*, see note on *si careat* in § 2.

XI. (X, 97)

1. **Actum, quem debuisti]** 'the right course', lit. which you ought (to have followed).

qui Christiani...] 'who had been brought before you as Christians'.

Neque enim...] 'for no rule can be laid down as a general law (*in universum*) to provide as it were a fixed standard', lit. nor can something be decided. *quod habeat* is a 'generic' subjunctive: 'such as to provide'.

2. **Conquirendi non sunt]** 'no search must be made for them'.

si deferantur...sunt] Note the 'mixed' condition: 'if they should be brought to trial and proved (guilty) they must be punished; with this proviso however (*ita tamen ut*, lit. in such a way that) that (any one) who denies...and proves his statement by his actions (*re ipsa*), that is, by..., although he has been suspected in the past, is to obtain pardon because of his repentance'. *negaverit* and *fecerit* are probably future perfect, though they might also be perfect subjunctive. *quamvis* is again used with a participle alone.

Sine auctore...] 'informations laid anonymously must not be admitted in any accusation; for this is both a very bad precedent and not in accordance with the spirit of our age'. *exempli* and *saeculi* are partitive genitives.

The Christians were first persecuted under Nero, when they were made scapegoats for the burning of Rome in A.D. 64, of which Nero accused them in order to avert suspicion from himself. Hundreds of them were burned in the *tunica molesta*, or 'shirt of pain', consisting of a coat of tar, clad in which the victims stood as living torches. It does not seem that there was any systematic persecution of Christians under the succeeding emperors, but the religion remained a forbidden one from the time of Domitian and any conscientious governor like Pliny could no doubt prosecute them whenever he wished. Christianity was the only illegal religion because its followers refused to worship the emperor's image, which meant treason; the Jews had been granted special privileges of exemption from emperor-worship.

VOCABULARY

Diphthongs, final *-i*'s and *-o*'s are long (except that in Silver Latin the final *-o* is often made short), and also vowels before two consonants (except that if *l* or *r* is the second consonant the syllable can sometimes be long or short, e.g. pătris); these long vowels are not marked. All other long vowels are marked, and any unmarked vowel can be assumed to be short, unless it comes under the rules just given.

Abbreviations

1, 2, 3 or 4 after a verb means that it is a regular verb like *amo, moneo, rego* or *audio.*

abl.	ablative.	indef.	indefinite.
acc.	accusative.	interj.	interjection.
adj.	adjective.	interrog.	interrogative.
adv.	adverb.	intr.	intransitive verb.
c.	common.	m.	masculine.
comp.	comparative.	n.	neuter.
conj.	conjunction.	p.p.p.	past participle passive.
dat.	dative.	part.	present participle.
defect.	defective.	pass.	passive.
dep.	deponent verb.	pl.	plural.
f.	feminine.	prep.	preposition.
gen.	genitive.	pron.	pronoun.
impers.	impersonal verb.	superl.	superlative.
indecl.	indeclinable.	tr.	transitive verb.

ā, ab, prep. with abl. *by, from*

abdo, -dere, -didi, -ditum, tr. *conceal*

abeo, -īre, -ii, -itum, intr. *go away, leave*

ābicio, -icere, -iēci, -iēctum, tr. *throw down*

abnego, 1, tr. *refuse, keep back*

abrumpo, -ere, -rūpi, -ruptum, tr. *break off*

abscēdo, -ere, -cessi, -cessum, intr. *depart*

abscondo, -ere, -didi, -ditum, tr. *hide*

ac, atque, conj. *and*

accēdo, -ere, -cessi, -cessum, intr. *approach, be added*

accido, -ere, -cidi, —, intr. usually impers. *happen*

accipio, -cipere, -cēpi, -ceptum, tr. *receive, take, hear*

accubo, -āre, —, —, intr. *recline at table*
ācer, ācris, ācre, adj. *sharp, eager, careful, of keen intellect, bracing*
Acīlius, -i, m. a friend of Pliny's
actum, -i, n. *deed, exploit*
actus, -ūs, m. *course, action*
acūtus, -a, -um, adj. *quick-witted*
ad, prep. with acc. *to, at, beside*
addo, -ere, -didi, -ditum, tr. *add, give*
addūco, 3, tr. *bring, lead on*
adeo, -īre, -ii, -itum, tr. and intr. *approach*
adeo, adv. *so much, so, to such an extent*
adfingo, -ere, -finxi, -fictum, tr. *add falsely, make up*
adfirmo, 1, tr. *declare, tell*
adhūc, adv. *still, as yet*
adiaceo, -ēre, -iacui, —, intr. *lie near*
ādicio, -icere, -iēci, -iectum, tr. *add, raise*
adiuvo, -āre, -iūvi, -iūtum, tr. *help*
adloquor, -i, -locūtus, dep. tr. *speak to, address*
adlūdo, -ere, -lūsi, -lūsum, intr. *play, play beside*
admitto, -ere, -mīsi, -missum, tr. *allow, admit*
admoneo, 2, tr. *warn, advise*
admoveo, -ēre, -mōvi, -mōtum, tr. *move towards*; in pass. *rise towards*
adnato, 1, intr. *swim up*
adnoto, 1, tr. *mark down, remark*
adpāreo, 2, intr. *appear*
adpropinquo, 1, intr. *approach, draw near*
adsero, -ere, -serui, -sertum, tr. *set free, claim*
adsiduus, -a, -um, adj. *constant*
adspicio, -spicere, -spexi, -spectum, tr. *look at, see*
adstruo, -ere, -struxi, -structum, tr. *add*
adsum, -esse, -fui, intr. *be present, come*
adsurgo, -ere, -surrexi, -surrectum, intr. *get up, arise*
adulterium, -i, n. *adultery* (also pl.)
advento, 1, intr. *approach*
adventus, -ūs, m. *arrival*
adversus, -a, -um, adj. *hostile, opposite*; n. pl. *adversity*
adversus, prep. with acc. *against*
advesperascit, -ere, -vesperāvit, —, intr., impers. *evening approaches*
aedēs, -ium, f. pl. *house*
aeger, -gra, -grum, adj. *sick*
aegrē, adv. *unwillingly, with difficulty*
aegrōto, 1, intr. *be ill*
Aeolus, -i, m. god of the winds
aequālis, -e, adj. *equal*; m. pl. *friends, contemporaries*
aequē, adv. *equally*
aequor, -oris, n. *sea*
aeripes, -pedis, adj. *brazen-footed*
aes, aeris, n. *bronze* (also pl.)
aestās, -ātis, f. *summer, summer heat*
aestimo, 1, tr. *value, take in*
aestuārium, -i, n. *estuary*
aestus, -ūs, m. *heat, tide*

aetās, -ātis, f. *age*
aeternitās, -ātis, f. *eternity, immortality*
aeternus, -a, -um, adj. *everlasting*
aethereus, -a, -um, adj. *of the upper air, heavenly*
aevum, -i, n. *time*
affero, -ferre, attuli, allātum, tr. *bring, bring near*
Āfrica, -ae, f. the province of *Africa*
Āfricānus, -i, m. the surname of the two Scipios
ager, agri, m. *field, land*
agito, 1, tr. *make plans, move*
agmen, agminis, n. *mass, crowd*
agnosco, -ere, -nōvi, -nitum, tr. *recognize*
ago, -ere, ēgi, actum, tr. *do, play, pass, move, plead* (a cause); **quid agit?** *how is he?*
agrestis, -is, m. *countryman, rustic*
āio, ait, defect. *say*
āla, -ae, f. *wing*
Albānus, -a, -um, adj. *of* or *at Alba*, a town near Rome
Albīnus, -i, m. a friend of Pliny's
albus, -a, -um, adj. *white*
Alcīdēs, -ae, m. grandson of Alceus, i.e. *Hercules*
Alcimus, -i, m. one of Martial's slaves
Alcinous, -oi, m. Homeric king of Phaeacia
ālea, -ae, f. *dice, gambling*
Algidus, -a, -um, adj. *of Algidum*, a hill-town in Latium
alibi, adv. *elsewhere*
aliēnus, -a, -um, adj. *belonging to another*
aliōqui, aliōquīn, adv. *in other ways, otherwise*
aliquando, adv. *sometimes, once*
aliquis, aliquid, indef. pron. *someone, some*
aliter, adv. *otherwise*
alius, -a, -ud, adj. *other, another, one...another, some ...others*
alo, -ere, -ui, altum, tr. *nourish, feed*
Alsiensis, -e, adj. *at Alsium*, a town in Etruria
alter, -era, -erum, adj. *the one or the other of two, another, second*
alternus, -a, -um, adj. *alternate*; **vice alternā**, *in turn*
altus, -a, -um, adj. *high, lofty, deep*; n. **altum**, *the sky, the open sea*; adv. **altē**, *from on high*; **altissimē**, *as far as possible*
Amāzon, -onis, acc. -ona, f. *an Amazon*
ambrosius, -a, -um, adj. *ambrosial, divine*
ambustus, -a, -um, p.p.p. *scorched*
āmentia, -ae, f. *madness*
amīcitia, -ae, f. *friendship*
amīcus, -i, m. *friend*; also adj. *friendly*
āmitto, -ere, -mīsi, -missum, tr. *lose*
amnis, -is, m. *river*
amo, 1, tr. *love, like*
amoenitās, -ātis, f. *pleasantness*
amor, amōris, m. *love*
amphitheātrum, -i, m. *amphitheatre*

amplector, -i, -plexus, dep. tr. *embrace, cover, seize*
amplitūdo, -inis, f. *size, corpulence*
amplius, comp. adv. *more*
amputo, 1, tr. *cut off*
an, interrog. conj. *or*; also = *num*
ancilla, -ae, f. *maid-servant*
angustus, -a, -um, adj. *narrow, obstructed, small*
anima, -ae, f. *life, breath*
animal, -ālis, n. *animal, creature*
animus, -i, m. *mind*
Anna Perenna, -ae, f. an old Italian goddess
annus, -i, m. *year*
ante, prep. with acc. *before*; adv. *before, first*
antīquus, -a, -um, adj. *ancient*
Antium, -i, n. a town in Latium
anus, -ūs, f. *old woman*
aper, apri, m. *boar*
aperio, -īre, aperui, apertum, tr. *open*; **in aperto**, *in the open air*
Apollināris, -is, m. a friend of Martial's
appello, 1, tr. *address, call upon*
Appia (via), -ae, f. *the Appian Way*
apto, 1, tr. *fit, make ready*
apud, prep. with acc. *among, before, with*
aqua, -ae, f. *water*
āra, -ae, f. *altar*
arātor, -ōris, m. *ploughman*
arbor, -oris, f. *tree*
Arcadius, -a, -um, adj. *Arcadian* (a district of Greece)
ardeo, -ēre, arsi, arsum, intr. *burn, glow*
ārea, -ae, f. *courtyard*
Argīlētum, -i, n.; also **Argi Lētum**, a district of Rome; adj. **Argīlētānus**, -a, -um
arguo, -ere, -ui, -ūtum, tr. *prove guilty*
argūtus, -a, -um, adj. *witty, chirping*
arma, -ōrum, n. pl. *arms*
Arpīnus, -a, -um, adj. *of Arpinum*, i.e. Ciceronian
Arria, -ae, f. the heroic wife of Paetus; also her daughter
ars, artis, f. *art*
arx, arcis, f. *citadel, height*
ascendo, -ere, -cendi, -censum, tr. *climb, embark on*
asper, aspera, asperum, adj. *rough, unfriendly*
aspicio, -spicere, -spexi, -spectum, tr. *see, observe*
astrum, astri, n. *star*; pl. *heaven*
astus, -ūs, m. *guile*
at, conj. *but*
āter, atra, atrum, adj. *black*
Atestānus, -i, m. a Roman lawyer
Athēnae, -ārum, f. pl. *Athens*
Athēnodōrus, -i, m. a Greek philosopher
Atilius, -i, m. Rūfus, governor of Syria
atque, ac, conj. *and*
Atrectus, -i, m. a bookseller at Rome
atrium, -i, n. *hall*
atrox, atrōcis, adj. *fierce, terrible*
attendo, -ere, -tendi, -tentum, tr. *listen*
attentē, adv. *closely*

attero, -ere, -trīvi, -trītum, tr. *exhaust*
Atthis, -idis, f. adj. *Athenian*
Atticus, -a, -um, adj. *Athenian*
attollo, -ere, —, —, tr. *magnify*
attonitus, -a, -um, p.p.p. *terrified*
auctor, -ōris, m. *author, creator, informant*
auctōritās, -ātis, f. *authority*
audācia, -ae, f. *boldness*
audeo, -ēre, ausus, semi-dep. intr. *dare*; part. **audens**, *bold*
audio, 4, tr. *hear*
aufero, -ferre, abstuli, ablātum, tr. *take away, rob*; in pass. *hurry off*
augeo, -ēre, auxi, auctum, tr. *increase*
auguror, 1, dep. tr. *foretell*
Augustus, -i, m. the emperor *Augustus*
Augustus, -a, -um, adj. *Augustan, imperial*
aura, -ae, f. *breeze*
aureus, -a, -um, adj. *golden*
auris, -is, f. *ear*
aurum, -i, n. *gold*
Ausonius, -a, -um, adj. *Italian*
aut, conj. *or, either...or*
autem, adv. *but, moreover*
auxilium, -i, n. *help*; pl. *auxiliaries*
Aventīnus, -a, -um, adj. *of the Aventine Hill* at Rome
avia, -ae, f. *grandmother*
Avītus, -i, m. Octāvius, legate in Africa
avunculus, -i, m. *uncle*

Baetis, -is, acc. -in, m. a river in Spain
Bāiānus, -a, -um, adj. *of Baiae*, a town on the coast of Latium
balineum, balneum, -i, n. *bath*
barba, -ae, f. *beard*
barbarus, -a, -um, adj. *foreign, barbarian*
bāsium, -i, n. *kiss*
Bassus, -i, m. a friend of Martial's
beātus, -a, -um, adj. *happy, blessed*
bellum, -i, n. *war*
bellus, -a, -um, adj. *pretty*
bene, adv. *well, fully*
Bilbilis, -is, f. a town in Spain where Martial was born
blaesus, -a, -um, adj. *lisping*
blandus, -a, -um, adj. *charming, tame, seductive, gentle*
bōlētus, -i, m. *mushroom*
bonus, -a, -um, adj. *good*; n. pl. **bona**, *goods, happiness*
bōs, bovis, c. *ox, cow*
Bōterdum, -i, n. a place in Spain
bractea, -ae, f. *gold-leaf, gold*
brevis, -e, adj. *short, tiny, shallow*
Britannus, -a, -um, adj. *British*; also m. plur. *Britons*
brūma, -ae, f. *winter*
buxus, -i, f. *box-tree*

C = Gāius, Roman proper name
Caeciliānus, -i, m. a profligate Roman
Caecīna, -ae, m. see **Paetus**
caecus, -a, -um, adj. *dark, blind*
caedēs, -is, f. *slaughter*
caedo, -ere, cecīdi, caesum, tr. *kill, break, cut off*

Caelius, -i, m. *the Caelian Hill* at Rome
caelum, -i, n. *climate, sky, air, position*
caeruleus, -a, -um, adj. *woad-painted, blue*
Caesar, -aris, m. *Caesar*
Caesariānus, -a, -um, adj. *of Caesar*
caespes, caespitis, m. *turf*
Cāiēta, -ae, f. a sea-coast town of Latium
calamus, -i, m. *fishing-rod*
calcio, 1, tr. *shoe*; in pass. *be shod*
calidus, -a, -um, adj. *warm, hot*
cālīgo, -inis, f. *vapour, darkness*
Calydōnius, -a, -um, adj. *Calydonian* (in Greece)
Camēna, -ae, f. *Muse*
Campānia, -ae, f. a district of Italy
campus, -i, m. *plain*
candidus, -a, -um, adj. *white*
candor, -ōris, m. *sincerity, frankness*
Canīnius, -i, m. a friend of Pliny's
canis, -is, c. *hound*
cānus, -a, -um, adj. *hoary, aged*
capax, -ācis, adj. *large, spacious*
capillus, -i, m. *hair*
capio, capere, cēpi, captum, tr. *take, receive, enjoy*
Capreae, -ārum, f. pl. the island of *Capri*
caput, capitis, n. *head, person*
careo, 2, intr. with abl. *lack, be free from*
carīna, -ae, f. *keel, ship*
carmen, carminis, n. *poem, hymn*
Carthāgo, -ginis, f. *Carthage*
Cārus, -i, m. Mettius, a notorious informer
cārus, -a, -um, adj. *dear, precious*
Castalis, -idis, f. adj. *Castalian*
castīgo, 1, tr. *upbraid*
Castor, -oris, acc. -ora, m. a god, the brother of Pollux
castus, -a, -um, adj. *chaste, pure*
cāsus, -ūs, m. *chance, fall, adventure, misfortune, destruction*
catella, -ae, f. *puppy*
catēna, -ae, f. *chain*
catēnātus, -a, -um, p.p.p. *chained up*
cathedra, -ae, f. *chair*
Catilīna, -ae, f. *Catiline*, a Roman revolutionary
Cato, -ōnis, m. a famous censor and moralist
Catullus, -i, m. a famous poet
causa, -ae, f. *cause, reason, case*
causidicus, -i, m. *lawyer, barrister*
cautē, adv. *cautiously*
caveo, -ēre, cāvi, cautum, tr. and intr. *beware, provide*
cēdo, -ere, cessi, cessum, intr. *give way, yield*
celebritās, -ātis, f. *fame*
celebro, 1, tr. *throng to, celebrate, make famous*
celeuma, -atis, n. *shout, order*
celsus, -a, -um, adj. *lofty*
Celtiber, -ēra, -ērum, adj. *Spanish, Celtiberian*
cēna, -ae, f. *dinner*
cēno, 1, intr. and tr. *dine, dine off*
censūra, -ae, f. *censorship*

centēnus, -a, -um, adj. *a hundred each, hundred*
centum, indecl. adj. *hundred*
centumviri, -ōrum, m. pl. *members of the Hundred Court*
cērae, -ārum, f. pl. *wax tablets*
Cerēs, Cereris, f. goddess of agriculture
cerno, -ere, crēvi, crētum, tr. *see*
certāmen, -minis, n. *contest*
certus, -a, -um, adj. *certain, fixed, resolved*; adv. **certē**, *at any rate*
cervīcal, -ālis, n. *pillow*
cervus, -i, m. *stag*
cesso, 1, intr. *hesitate*
cēterus, -a, -um, adj. (masc. nom. sing. not used), *the rest, the other*
charta, -ae, f. *paper, page*
Christiānus, -a, -um, adj. *Christian*; also m. noun
Christus, -i, m. *Christ*
cibus, -i, m. *food*
cicāda, -ae, f. *grasshopper*
Cicero, -ōnis, m. M. Tullius, a famous Roman orator
cingo, -ere, cinxi, cinctum, tr. *surround*
cinis, cineris, m. *ash, cinder*
circā, prep. with acc. *around, round*; also adv.
Circē, Circēs, f. a witch living at Circeii
circensēs, -ium, m. pl. *public games*
circumactus, -a, -um, p.p.p. *borne round*
circumeo, -īre, -ii, -itum, tr. *go round*
circumfero, -ferre, -tuli, -lātum, tr. *attend, accompany*
circumiacens, -ntis, part. *surrounding, neighbouring*
circumsisto, -ere, -steti, —, tr. *surround*
cito, citius, adv. *quickly*
cito, 1, tr. *summon, carry at full speed*
cīvīlis, -e, adj. *civil*
Cīvis, -is, m. a Roman lawyer
cīvis, -is, c. *citizen*
cīvitās, -ātis, f. *state, city*
clādēs, -is, f. *disaster*
clāmo, 1, intr. *shout*
clāmor, -ōris, m. *shout, cry*
clāmōsus, -a, -um, adj. *noisy*
clāritās, -ātis, f. *fame, brightness*
clārus, -a, -um, adj. *gleaming, bright, famous*
classis, -is, f. *fleet*
Claudia, -ae, f. Rūfīna, a British girl
Claudius, -i, m. the emperor *Claudius*
claudo, -ere, clausi, clausum, tr. *shut, close, block up*
clīvus, -i, m. *slope, hill*
cochlea, -ae, f. *snail shell*
cocus, -i, m. *cook*
cōdicilli, -ōrum, m. pl. *a letter*
coeo, -īre, -ii, -itum, intr. *come together, flock together*
coepi, -isse, defect. tr. *began*
cognitio, -ōnis, f. *enquiry, trial*
cognosco, -noscere, -nōvi, -nitum, tr. *learn, listen to*
cōgo, -ere, coēgi, coactum, tr. *compel*
cohaereo, -ēre, -haesi, -haesum, intr. *be connected with*
cohibeo, 2, tr. *restrain*
collātio, -ōnis, f. *comparison*

colligo, -ere, -lēgi, -lectum, tr. *collect, infer*
collis, -is, m. *hill*
collum, -i, n. *neck*
collūsor, -ōris, m. *playmate*
colo, -ere, colui, cultum, tr. *worship, dwell in, adorn, haunt, feel, court, feed*
colōnia, -ae, f. *colony*
colōnus, -i, m. *husbandman, farmer*
colossus, -i, m. *Colossus*, i.e. giant statue
columba, -ae, f. *dove*
columna, -ae, f. *column*
coma, -ae, f. *hair, foliage*
comedo, -ere, -ēdi, -ēsum, tr. *eat*
comes, comitis, m. *companion*
cōmis, -e, adj. *friendly*
cōmissātor, -ōris, m. *reveller*
comitor, 1, dep. tr. *accompany*
commendo, 1, tr. *introduce, entrust, put at one's disposal*
committo, -ere, -mīsi, -missum, tr. *commit*; with **ut**, *act in such a way that*
commodo, 1, tr. *lend*
commodus, -a, -um, adj. *comfortable*
commūnis, -e, adj. *common*; **in commūne,** *together*
comparo, 1, tr. *compare*
compēs, -pedis, f. *fetter*
complector, 3, dep. tr. *embrace*
compōno, -ere, -posui, -positum, tr. *compose, write*; p.p.p. **compositus,** *calm*
comprehendo, -ere, -prehendi, -prehensum, tr. *arrest*
concerpo, -ere, -cerpsi, -cerptum, tr. *pluck*
concido, -ere, -cidi, —, intr. *fall down*
concordia, -ae, f. *concord*
concurro, -ere, -curri, -cursum, intr. *run together, fight*
concutio, -cutere, -cussi, -cussum, tr. *shake*
conditōrium, -i, n. *tomb*
condo, -ere, -didi, -ditum, tr. *bury, compose, store*
condūco, 3, tr. *hire, rent*
confectus, -a, -um, p.p.p. *worn out*
confero, -ferre, -tuli, -lātum, tr. *collect*; **sē conferre,** *go*
confirmo, 1, tr. *confirm*
confiteor, -ēri, -fessus, dep. tr. *admit, confess, show*
confluo, -ere, -fluxi, —, intr. *flock together*
cōniecto, 1, tr. *guess*
cōniunx, -iugis, c. *wife, husband*
conquīro, -ere, -quīsīvi, -quīsītum, tr. *search out*
consecro, 1, tr. *consecrate*
consīdero, 1, tr. *consider*
consīdo, -ere, -sēdi, -sessum, intr. *stop*
consilium, -i, n. *plan*
consisto, -ere, -stiti, -stitum, intr. *halt*
consōlor, 1, dep. tr. *comfort*
conspicio, -spicere, -spexi, -spectum, tr. *see*
conspicuus, -a, -um, adj. *obvious*
constantia, -ae, f. *courage*
constat, -āre, -stitit, impers. *it is well known*
constituo, -ere, -ui, -ūtum, tr. *determine*
constringo, -ere, -strinxi, -strictum, tr. *fasten*

consuēsco, -suēscere, -suēvi, -suētum, intr. *become used, grow accustomed*
consulāris, -e, adj. *consular*
consulo, -ere, -sului, -sultum, tr. and intr. *consult, take counsel*; with dat. *consider*
consultātio, -ōnis, f. *consultation*
consulto, 1, intr. *take counsel*
consūmo, -ere, -sumpsi, -sumptum, tr. *eat up*
consurgo, -ere, -surrexi, -surrectum, intr. *rise up*
contāgio, -ōnis, f. *infection*
contero, -ere, -trīvi, -trītum, tr. *wear out*
contexo, -ere, -texui, -textum, tr. *add, write*
contineo, -ēre, -tinui, -tentum, tr. *contain*
contingo, -ere, -tigi, -tactum, tr. *come across*; impers. *fall to one's lot*
contrā, prep. with acc. *against, opposite to*; adv. *on the other hand*
contrārius, -a, -um, adj. *contrary, opposite*
contumēlia, -ae, f. *insult*
contundo, -ere, -tudi, -tūsum, tr. *break, strike*
convenio, -īre, -vēni, -ventum, intr. *come together, meet*
converto, -ere, -verti, -versum, tr. *turn*
convictus, -ūs, m. *company, social intercourse*
cōpia, -ae, f. *eloquence*; **cōpiam facere**, *give a chance of, put at one's disposal*
cōpo, -ōnis, m. *tavern-keeper*
cōr, cordis, n. *heart*
Corduba, -ae, f. a town in Spain
Cornēlius, -i, m. see **Priscus** and Tacitus
cornū, -ūs, n. *horn*
corōna, -ae, f. *garland*
corōno, 1, tr. *crown*
corpus, corporis, n. *body*
corrigo, -ere, -rexi, -rectum, tr. *reform, cure*
corripio, -ripere, -ripui, -reptum, tr. *upbraid*
corruo, -ere, -rui, —, intr. *collapse*
Corybās, -ntis, m. *Corybant*, priest of Cybele
Cosmus, -i, m. a perfumer at Rome
crassus, -a, -um, adj. *uncouth, thick*
crēber, -bra, -brum, adj. *frequent*; adv. **crēbro**, *frequently*
crēbresco, -ere, crēbrui, —, intr. *increase*
crēdo, -ere, -didi, -ditum, tr. and intr. *believe, suppose*
creo, 1, tr. *produce*
cresco, -ere, crēvi, crētum, intr. *grow, increase*; part. **crescens**, *youthful*
crīmen, crīminis, n. *charge, accusation*
cristātus, -a, -um, adj. *crested*
crūdēlis, -e, adj. *cruel*
crūdus, -a, -um, adj. *raw, green* (of pepper)
cruor, -ōris, m. *blood*
crūs, crūris, n. *leg*
cubīc(u)lum, -i, n. *bed-room*
cubīle, -is, n. *bed*
cubo, -āre, -ui, -itum, intr. *lie*
cucumis, -is, m. *cucumber*
culex, -icis, m. *gnat*
culmen, -minis, n. *roof-top*
culpa, -ae, f *guilt*

culter, -tri, m. *knife*
cum, prep. with abl. *with*
cum, conj. *when, since, although*
cunctātio, -ōnis, f. *hesitation*
cunctor, 1, dep. *delay, hesitate*
cupio, cupere, -īvi or -ii, -ītum, tr. *desire*
cūr, adj. *why?*
cūra, -ae, f. *care, anxiety, duty*
cūro, 1, tr. *look after, refresh*
curro, -ere, cucurri, cursum, intr. *run, speed*
cursus, -ūs, m. *course, speed*
Curtius, -i, m. Rūfus, governor of Africa
curvātus, -a, -um, p.p.p. *curved*
curvus, -a, -um, adj. *curved*
custōdio, 4, tr. *guard, look after*
Cȳaneus, -a, -um, *of the Cyanean rocks*
Cybelē, -ēs, f. the goddess *Cybele*

damno, 1, tr. *condemn*
Daphnis, -nidis, m. name of a Roman
daps, dapis, f. *feast*
Dardanis, -idis, f. adj. *Trojan*
dē, prep. with abl. *concerning, about, from, out of*
dea, -ae, f. *goddess*
dēbeo, 2, tr. *owe, ought*
dēcēdo, -ere, -cessi, -cessum, intr. *depart, die*
December, -bris, m. the month *December*
decimus, -a, -um, adj. *tenth*
dēcurro, -ere, -cucurri and -curri, -cursum, intr. *hasten*
decus, decoris, n. *glory, grace*
dēdisco, -ere, -didici, —, tr. *unlearn, forget*
dēdo, -ere, -didi, -ditum, tr. *hand over*
dēdūco, 3, tr. *launch*
dēfero, -ferre, -tuli, -lātum, tr. *carry, bring to trial*
dēficio, -ficere, -fēci, -fectum, intr. *be wanting, be eclipsed*
dēflecto, -ere, -flexi, -flexum, tr. and intr. *turn aside*
dēfleo, -ēre, -flēvi, -flētum, tr. *bewail*
dēfungor, -i, -functus, dep. *die*
dehisco, -ere, —, —, intr. *split*
dēicio, -icere, -iēci, -iectum, tr. *throw down, scatter*
dein, **deinde**, adv. *then, next*
dēlicātus, -a, -um, adj. *dainty, graceful*
dēliciae, -ārum, f. pl. *darling*
delphīnus, -i, m. *dolphin*
dēmonstro, 1, tr. *point out*
Dēmosthenēs, -is, m. a famous Greek orator
dēmum, adv. *at last*
dēnārius, -i, m. a silver coin, *denarius*
dēni, -ae, -a, pl. adj. *ten each, ten*
dēnique, adv. *in fact, finally*
dens, dentis, m. *tooth*
densus, -a, -um, adj. *thick*
dēpello, -ere, -puli, -pulsum, tr. *drive away, avoid, escape*
dēpōno, -ere, -posui, -positum, tr. *put down*; n. noun, **dēpositum**, *a deposit*
dēposco, -ere, -poposci, —, tr. *demand*
dēprecor, 1, dep. tr. *beg, pray*
dēprehendo, -ere, -prehendi, -prehensum, tr. *notice*

dērīsor, -ōris, m. *mocker, mimic*
descendo, -ere, -scendi, -scensum, intr. *descend, come down*
dēsero, -ere, -serui, -sertum, tr. *desert*
dēsīderium, -i, n. *longing*
dēsīdero, 1, tr. *long for, miss*
dēsino, -ere, -sīvi or -sii, -situm, tr. and intr. *cease*
dēsōlātus, -a, -um, p.p.p. *deserted*
despēro, 1, intr. *despair*
destituo, -ere, -ui, -ūtum, tr. *abandon*
dēsum, -esse, -fui, intr. *be lacking*
dēterior, -ius, comp. adj. *worse, more deadly*
dētineo, -ēre, -tinui, -tentum, tr. *keep, keep back*
dētondeo, -ēre, -tondi, -tonsum, tr. *cut the hair of*
deus, -i, m. *god*
dextera, **dextra**, -ae, f. *right hand*
diaeta, -ae, f. *living-room*
Diāna, -ae, f. the goddess of the moon and of hunting
dīco, 3, tr. *say, call, sing*; **male dīco**, *curse*; n. noun, **dictum**, *word, saying, wit*
dictito, 1, tr. *keep saying*
dicto, 1, tr. *keep saying, dictate*
diēs, diēi, m. *day*
differo, -ferre, distuli, dīlātum, tr. *put off*; intr. *differ*
difficultās, -ātis, f. *difficulty*
diffugio, -fugere, -fūgi, —, intr. *flee in all directions*
diffundo, -ere, -fūdi, -fūsum, tr. *spread*
digitus, -i, m. *finger*
dignus, -a, -um, adj. *worthy*
dīlābor, -i, -lapsus, dep. *vanish*
dīligo, -ere, -lexi, -lectum, tr. *love*
dīmidius, -a, -um, adj. *half*
dīmitto, -ere, -mīsi, -missum, tr. *send away*
dīrimo, -ere, -rēmi, -remptum, tr. *separate*
dīrus, -a, -um, adj. *dread, terrible*
discēdo, -ere, -cessi, -cessum, intr. *depart, vanish*
discingo, -ere, -cinxi, -cinctum, tr. *ungird, strip*
discipulus, -i, m. *pupil*
disco, -ere, didici, —, tr. *learn, hear*
discrīmen, -crīminis, n. *peril, difference*
discursus, -ūs, m. *running about*, hence, *flash*
dīsertus, -a, -um, adj. *eloquent*
dispenso, 1, tr. *give out*
disputo, 1, intr. *argue*
diū, adv. *for a long time*
dīus, see dīvus
dīves, dīvitis, adj. *rich*
dīvido, -ere, -vīsi, -vīsum, tr. *divide, shut off*
dīvīnus, -a, -um, adj. *divine, inspired*
dīvus, -a, -um, adj., also **dīus**, *divine*; **sub dīo**, *in the open air*
do, dare, dedi, datum, tr. *give, surrender*
doceo, -ēre, -ui, doctum, tr. *teach, tell, prove*; p.p.p. **doctus**, *learned*
doleo, 2, intr. *hurt*; tr. *grieve for*
dolor, -ōris, m. *grief*
dolus, -i, m. *treachery, hidden danger*

domina, -ae, f. *queen*
dominus, -i, m. *master, lord*; as adj. *imperial*
Domitiānus, -i, m. the emperor *Domitian*
domo, -āre, -ui, -itum, tr. *tame, subdue, cultivate*
domus, -ūs, f. *house, home*
dōnec, conj. *until*
dōno, 1, tr. *give*
dormio, 4, intr. *sleep*
Dryas, -adis, f. *wood-nymph, Dryad*
dubito, 1, intr. *doubt*
dubius, -a, -um, adj. *doubtful, anxious*
dūco, 3, tr. *lead, draw out, celebrate, induce*
dulcis, -e, adj. *sweet, pleasant*
dum, conj. *while, until*
duo, duae, duo, pl. adj. *two*
duodēvīcensimus, -a, -um, adj. *eighteenth*
dūrus, -a, -um, adj. *hard*
dux, ducis, m. *general, emperor*

ē, ex, prep. with abl. *from*
ēbrius, -a, -um, adj. *drunken*
ecce, interj. *see, behold*
ecquid, interrog. adv. *at all*
ēdictum, -i, n. *decree*
ēdo, -ere, ēdidi, ēditum, tr. *produce*; p.p.p. **ēditus**, *sprung from*
ēdūco, 3, tr. *drag out*
effero, -ferre, extuli, ēlātum, tr. *carry out*; in pass. *rise*
efficio, -ficere, -fēci, -fectum, tr. *cause, bring about*
effigiēs, -ēi, f. *form, phantom*
effodio, -fodere, -fōdi, -fossum, tr. *dig up*
effulgeo, -ēre, -fulsi, —, intr. *shine out*
effundo, -ere, -fūdi, -fūsum, tr. *pour out*; **effūso cursū**, *at full speed*
ego, mē, mei, pron. *I*
ēgredior, -i, -gressus, dep. intr. *go out, go past, disembark*
ēgregius, -a, -um, adj. *illustrious, excellent*
ēligo, -ere, -lēgi, -lectum, tr. *choose*
ēmendo, 1, tr. *correct*
ēmergo, -ere, ēmersi, -sum, intr. *come out, flow out*
emo, -ere, ēmi, emptum, tr. *buy*
ēmoveo, -ēre, -mōvi, -mōtum, tr. *move, shift*
emptor, -ōris, m. *buyer*
enim, conj. *for*
ēnoto, 1, tr. *make notes on*
eo, īre, īvi or ii, itum, intr. *go*
eo, adv. *for that reason*
epigramma, -atis (gen. pl. -ōn), n. *epigram*
epistula, -ae, f. *letter*
eques, equitis, m. *knight*
equidem, adv. *I indeed*
equus, -i, m. *horse*
ergo, adv. *therefore*
ēripio, -ripere, -ripui, -reptum, tr. *snatch away, save*
Erōtion, Erōtii, f. and n. a slave-girl of Martial's
erro, 1, intr. *wander, be wrong*
error, -ōris, m. *mistake*
ērudītio, -ōnis, f. *learning*
ērudītus, -a, -um, p.p.p. *learned*
Esquiliae, -ārum, f. pl. *the Esquiline Hill* at Rome
essedum, -i, n. *carriage*
et, conj. *and, also, even, both... and*
etiam, conj. *also, even*

etsi, conj. *although*
Euphēmus, -i, m. Domitian's table-steward
ēvado, -ere, -vāsi, -vāsum, tr. and intr. *escape*
ēveho, -ere, -vexi, -vectum, tr. *lift up*
ēvenit, -īre, -vēnit, -ventum, impers. *it happens*
ex, ē, prep. with abl. *from, out of*
exanimis, -e, adj. *dead*
excēdo, -ere, -cessi, -cessum, intr. *depart*
excelsus, -a, -um, adj. *lofty*
excerpo, -ere, -cerpsi, -cerptum, tr. *select, make extracts from*
excido, -ere, -cidi, —, intr. *fall out, fall from one's lips*
excipio, -cipere, -cēpi, -ceptum, tr. *receive, meet*
excito, 1, tr. *rouse, increase*
exclāmo, 1, tr. *exclaim*
excūso, 1, tr. *excuse*
excutio, -cutere, -cussi, -cussum, tr. *shake out* or *off, pull tight, examine*
exemplum, -i, n. *example, precedent*
exequiae, -ārum, f. pl. *funeral rites*
exerceo, 2, tr. *keep busy, employ*
exēsus, -a, -um, p.p.p. *eaten away, fine*
exigo, -ere, -ēgi, -actum, tr. *finish, demand, pass*
exiguus, -a, -um, adj. *very small, tiny*
exilio, -īre, -ui, —, intr. *leap up*
eximius, -a, -um, adj. *remarkable*
existimo, 1, tr. *think*
exitium, -i, n. *death*
exitus, -ūs, m. *end, way out, outcome, death*
exolesco, -ere, -olēvi, -olētum, intr. *die out*
exorno, 1, tr. *adorn*
expecto, 1, tr. and intr. *wait, expect*
expedio, 4, tr. *unwind*
experīmentum, -i, n. *experience*
experior, -īri, -pertus, dep. *try, experience*
expōno, -ere, -posui, -positum, tr. *relate*
exprimo, -ere, -pressi, -pressum, tr. *represent, express*
extendo, -ere, -tendi, -tentum, tr. *extend*
exterreo, 2, tr. *terrify*
extinguo, -ere, -tinxi, -tinctum, tr. *quench, put out*
extraho, -ere, -traxi, -tractum, tr. *withdraw, drag out*
extruo, -ere, -truxi, -tructum, tr. *build, heap up*
exuo, -ere, -ui, -ūtum, tr. *put off*

faber, -bri, m. *smith*
Fabius, -i, m. a Roman name
fābula, -ae, f. *tale, story*
faciēs, -iēi, f. *face*
facilis, -e, adj. *easy, favouring, pliant*
facinus, -oris, n. *crime*
facio, facere, fēci, factum, tr. *do, make*
factum, -i, n. *deed*
facultās, -ātis, f. *opportunity*
Fadius, -i, m. see Rūfīnus
fallo, -ere, fefelli, falsum, tr. *deceive*; **fidem fallere**, *break one's word*
falsus, -a, -um, adj. *false*; adv. **falso**, *falsely*

fāma, -ae, f. *fame, reputation, rumour, glory*
famēs, -is, f. *hunger*
Fannia, -ae, f. daughter of the younger Arria
fastīdium, -i, n. *pride, contempt*
fastus, -ūs, m. *pride*
fateor, -ēri, fassus, dep. *confess*
fatīgo, 1, tr. *weary*
faucēs, -ium, f. pl. *throat*
Faunus, -i, m. *a Faun*
faveo, -ēre, fāvi, fautum, intr. *favour, support*
fax, facis, f. *torch*
fēcundus, -a, -um, adj. *fruitful, prolific*
fel, fellis, n. *gall, satire*
fēlix, -īcis, adj. *happy, fortunate, auspicious*
fēmina, -ae, f. *woman*
fenestra, -ae, f. *window*
fera, -ae, f. *wild beast*
ferē, adv. *about*
fēriātus, -a, -um, adj. *holiday*
feritās, -ātis, f. *fierceness*
fero, ferre, tuli, lātum, tr. *endure, bear, carry, take away*
ferrum, -i, n. *iron, sword, axe*
fervens, -ntis, part. *burning, hot*
fessus, -a, -um, adj. *weary*
fictus, -a, -um, p.p.p. *false, imitation*
fidēlis, -e, adj. *faithful*
Fidēnae, -ārum, f. pl. a town in Latium
fidēs, -ei, f. *faith, loyalty, truth, belief*
fīdūcia, -ae, f. *confidence*
figūra, -ae, f. *form, shape, phase*
fīlia, -ae, f. *daughter*
fīlius, -i, m. *son*
fingo, -ere, finxi, fictum, tr. *make up, imagine*
fīnis, -is, m. *end*; pl. *lands*
fīo, fieri, factus, intr. and pass. *become, be made, happen*
fistula, -ae, f. *pipe*
Flaccilla, -ae, f. possibly Martial's mother
flāgitium, -i, n. *crime*
Flāminia (via), -ae, f. *the Flaminian Way*, running north from Rome
flamma, -ae, f. *flame, torch*
Flāvus, -i, m. a friend of Martial's
flāvus, -a, -um, adj. *yellow, golden*
flecto, -ere, flexi, flexum, tr. and intr. *turn*
flūmen, flūminis, n. *river*
focilo, 1, tr. *revive*
focus, -i, m. *hearth, fire*
folium, -i, n. *leaf*
forīs, adv. *outside*
forma, -ae, f. *form, shape, standard*
Formiae, -ārum, f. pl. town in Latium; **Formiānus**, -a, -um, adj. *at Formiae*
formīca, -ae, f. *ant*
formīdo, -inis, f. *fear, peril*
formīdulōsus, -a, -um, adj. *terrible*
forsan, **forsitan**, adv. *perhaps*
fortasse, adv. *perhaps*
forte, adv. *by chance*
fortis, -e, adj. *brave, strong*
fortūna, -ae, f. *fortune*
forum, -i, n. *market, forum*
fossor, -ōris, m. *digger*
fragmentum, -i, n. *fragment*
fragor, -ōris, m. *noise*
frango, -ere, frēgi, fractum, tr. *break, disturb*

frāter, -tris, m. *brother*
frequens, -ntis, adj. *crowded*; adv. **frequenter**, *often*
fretum, -i, n. *sea*
frīgidus, -a, -um, adj. *chilly, cold*
frīgus, -oris, n. *coolness, cold*
fritillus, -i, m. *dice-box*
frons, frondis, f. *leaf, foliage*
frons, frontis, f. *forehead, face, front*
Fronto, -ōnis, m. possibly Martial's father
fructus, -ūs, m. *fruit, reward*
fruor, -i, fructus or fruitus, dep. intr. with abl. *enjoy*
fuga, -ae, f. *flight*
fugio, -ere, fūgi, —, tr. and intr. *flee, flee from*
fulcio, -īre, fulsi, fultum, tr. *support, stop*
fulgeo, -ēre, fulsi, —, intr. *gleam, flash*
fulgor, -ōris, m. *brightness, flash of lightning, gleam*
fūmus, -i, m. *smoke*
fūnus, -eris, n. *funeral, death*
furiālis, -e, adj. *mad, fearful*
furo, -ere, -ui, —, intr. *rage*
furor, -ōris, m. *madness*
furtum, -i, n. *theft*
futūrus, -a, -um, part. *future*; n. noun, *the future*

gallus, -i, m. *cock*
garrio, 4, intr. *prattle*
garrulus, -a, -um, adj. *talkative, loud-voiced*
gaudeo, -ēre, gāvīsus, semi-dep. intr. *rejoice (in)*
gaudium, -i, n. *joy*
gelidus, -a, -um, adj. *cool, chilly*
gemitus, -ūs, m. *groan*
gener, -eri, m. *son-in-law*
genetrix, -trīcis, f. *mother*
gens, gentis, f. *race*
genus, generis, n. *race, class*
Germānicus, -i, m. title taken by Domitian
gero, -ere, gessi, gestum, tr. *do, hold, wear, bear, carry on, have*; p.p.p. **gesta**, *exploits*
gestātor, -ōris, m. *traveller, carrier*
Geticus, -a, -um, adj. *Getic* (a tribe on the Danube)
gladius, -i, m. *sword*
glōria, -ae, f. *glory, fame*
glōrior, 1, dep. intr. *boast*
gradus, -ūs, m. *stair, step, degree*
grandis, -e, adj. *large, loud*
grātia, -ae, f. *thanks*; **grātiam agere, referre**, *thank*
gravis, -e, adj. *heavy, loud, serious*; adv. **graviter**
gravitās, -ātis, f. *dignity*
gremium, -i, n. *lap, arms*
gressus, -ūs, m. *step*
gubernāculum, -i, n. *helm* (also pl.)
gubernātor, -ōris, m. *helmsman*
gusto, 1, tr. and intr. *taste have a meal*

habēna, -ae, f. *rein*
habeo, 2, tr. *have, consider*
habito, 1, tr. and intr. *dwell in, dwell*
habitus, -ūs, m. *appearance*
hāc, adv. *by this way*
hactenus, adv. *so far*
haereo, -ēre, haesi, haesum, intr. *cling, accompany*
haesito, 1, intr. *hesitate*

hāmus, -i, m. *fish-hook*
Hannibal, -alis, m. famous Carthaginian general
harēna, -ae, f. *sand, arena*
harundo, -dinis, f. *reed, pen*
haud, adv. *not*
haurio, -īre, hausi, haustum, tr. *drain, drink*
haveo, -ēre, —, —, intr. *hail*
helciārius, -i, m. *bargee*
herba, -ae, f. *grass*
Herculēs, -is, m. the demi-god *Hercules*
Herculeus, -a, -um, adj. *of Hercules*
hērēs, -ēdis, c. *heir*
Hesperides, -um, f. pl. *the Hesperides,* daughters of Hesperus
Hesperius, -a, -um, adj. *western*
hesternus, -a, -um, adj. *of yesterday*
hetaeria, -ae, f. *society*
heu, interj. *alas*
hiātus, -ūs, m. *hole, gap, jaws*
hīc, haec, hōc, pron. *this, he, she, it*; adv. **hīc**, *here*
hilaris, -e, adj. *cheerful*
hinc, adv. *from here, from this*; **hinc atque hinc, hinc et inde,** *on all sides*
hio, 1, intr. *gape open*
Hippōnensis, -e, *of Hippo,* in N. Africa
hirundinīnus, -a, -um, adj. *of a swallow*
Hispānia, -ae, f. *Spain*
Hispānus, -a, -um, adj. *Spanish*
Hister, -tri, m. *the Danube*
historia, -ae, f. *history*
homo, hominis, m. *man*
honor, -ōris, m. *honour, office*
hōra, -ae, f. *hour*
horreo, -ēre, -ui, —, intr. *shiver, shrink*; part. **horrens**, *bristling, shaggy*; **horrendus**, *dreadful*
horresco, -ere, horrui, —, tr. *tremble at*
hortor, 1, dep. tr. *encourage*
hortus, -i, m. *garden*
hospes, -pitis, m. *stranger, guest*
hospitālis, -e, adj. *hospitable*
hospitālitās, -ātis, f. *hospitality*
hūc, adv. *hither*
hūmānus, -a, -um, adj. *human*
humus, -i, f. *soil, ground*
hȳdra, -ae, f. *water-snake*
Hylās, -ae, m. a beautiful youth
Hyperboreus, -a, -um, adj. *northern*

iaceo, 2, intr. *lie, lie down*
iacio, iacere, iēci, iactum, tr. *throw*
iacto, 1, tr. *throw*; part. **iactans**, *boastful*
iam, adv. *already, now*
Iāniculum, -i, n. a hill a mile west of Rome
iānitor, -ōris, m. *doorkeeper*
iānua, -ae, f. *door*
ibi, adv. *there*
īdem, eadem, idem, adj. *same*
identidem, adv. *again and again*
ideo, adv. *therefore*
īdōlon, -i, n. *ghost*
igitur, adv. *therefore*
ignārus, -a, -um, adj. *ignorant*
igneus, -a, -um, adj. *fiery*
ignis, -is, m. *fire*
ignōrantia, -ae, f. *ignorance*
ignōro, 1, tr. *not to know*

ignosco, -ere, -nōvi, -nōtum, intr. with dat. *pardon*
ignōtus, -a, -um, adj. *unknown*
ilicētum, -i, n. *oak-coppice*
ille, illa, illud, pron. *that, he, she, it*; adv. **illic**, *there*; **illinc**, *from there*; **illūc**, *thither, to that place*
Illyricum, -i, n. *Illyria*, a country N.W. of Greece
imāgo, -inis, f. *statue, phantom, appearance, form*
immineo, -ēre, —, —, intr. *overhang, threaten*
immo, adv. *in fact*
impendo, -ere, -pendi, -pensum, tr. *expend, employ*
imperātor, -ōris, m. *emperor*
imperium, -i, n. *supreme command, empire*
impero, 1, tr. and intr. *order*
impetro, 1, tr. and intr. *obtain, obtain a request*
impetus, -ūs, m. *force, rush*
impius, -a, -um, adj. *wicked, sacrilegious*
implico, -āre, -ui, -icitum, tr. *wind, wind round, seize*
improbus, -a, -um, adj. *shameful, shamefully long*
imprūdentia, -ae, f. *folly*
in, prep. with acc. *to, into, against, for*; with abl. *in, on*
inānis, -e, adj. *imaginary*
incalesco, -ere, -calui, —, intr. *grow hot*
incendium, -i, n. *fire*
incertus, -a, -um, adj. *uncertain*
incido, -ere, -cidi, -cāsum, intr. *fall down, fall upon, come across, come to one's notice*
incipio, -cipere, -cēpi, -ceptum, tr. *begin*
inclīno, 1, tr. *incline*; **inclīnāto diē**, *just after midday*
incoho, 1, tr. *begin*
incolo, -ere, -ui, -cultum, tr. *inhabit, live in*
incrēdibilis, -e, adj. *unbelievable*
incumbo, -ere, -cubui, -cubitum, intr. *lean over*
incūs, -cūdis, f. *anvil*
incūso, 1, tr. *blame*
inde, adv. *then, next, from there*
index, -dicis, c. *informer*
indicium, -i, n. *evidence, sign*
indico, 1, tr. *point out*
Indicus, -a, -um, adj. *Indian*
indignātio, -ōnis, f. *indignation*
indignus, -a, -um, adj. *disgraceful, unworthy*
induo, -ere, -ui, -ūtum, tr. *dress*
ineptus, -a, -um, adj. *unsuitable*
inerro, 1, intr. with dat. *wander before*
inertia, -ae, f. *slackness*
infāmis, -e, adj. *of ill repute*
infans, -ntis, c. *infant, child*
infantia, -ae, f. *infancy*
infēlix, -īcis, adj. *unhappy, ill-starred*
infero, -ferre, -tuli, -lātum, tr. *bring in*; in pass. *flow into*
inflexibilis, -e, adj. *inflexible*
infundo, -ere, -fūdi, -fūsum, tr. *pour in, pour over*
ingeniōsus, -a, -um, adj. *talented*
ingenium, -i, n. *genius, talent*
ingens, -ntis, adj. *huge*
ingenuus, -a -um, adj. *of a freeman*

ingrātus, -a, -um, adj. *ungrateful*
ingredior, -i, -gressus, dep. tr. *enter, reach*
inhabito, 1, tr. *dwell in*
initium, -i, n. *beginning*; **initio**, *at first*
iniungo, -ere, -iunxi, -iunctum, tr. *impose*
iniūria, -ae, f. *insult, wrong*
inlaesus, -a, -um, adj. *unharmed*
inlitterātus, -a, -um, adj. *uneducated*
inlūcescit, -ere, -luxit, impers. *day breaks*
inlūnis, -e, adj. *moonless*
inmōbilis, -e, adj. *motionless*
inmodicus, -a, -um, adj. *unrestrained*
inmortālis, -e, adj. *everlasting*
innītor, -i, -nixus, dep. intr. *lean on*, with dat.
innocens, -ntis, adj. *innocent*
innocuus, -a, -um, adj. *harmless, innocent*
innoxius, -a, -um, adj. *innocent*
innuo, -ere, -ui, -ūtum, intr. *beckon*
innutrītus, -a, -um, p.p.p. *brought up in*
inpello, -ere, -puli, -pulsum, tr. *set in motion, move forward*
inperfectus, -a, -um, adj. *unfinished*
inpingo, -ere, -pēgi, -pactum, tr. *strike against, strike*
inpleo, -ēre, -ēvi, -ētum, tr. *fill, complete, fulfil*
inpōno, -ere, -posui, -positum, tr. *place upon, put on board*
inprīmīs, adv. *particularly, in particular*
inputo, 1, tr. *put down to one's account*
inquam, inquit, defect. *say*
inquiētus, -a, -um, adj. *anxious, disturbed, restless*
inrumpo, -ere, -rūpi, -ruptum, tr. *burst into*
insānia, -ae, f. *madness*
insānio, 4, intr. *be mad*
inscrībo, -ere, -scripsi, -scriptum, tr. *inscribe*
insero, -ere, -ui, -sertum, tr. *thrust in, throw in, wrap round*
insilio, -īre, -ui, —, intr. *leap upon*
insono, -āre, -ui, —, intr. *make a noise*
instans, -ntis, adj. *pressing*; comp. adv. **instantius**, *more pressingly*
institor, -ōris, m. *shop-keeper*
instruo, -ere, -struxi, -structum, tr. *instruct*
integer, -gra, -grum, adj. *honourable, unhurt*
intendo, -ere, -tendi, -tentum and -tensum, tr. *plan, apply*
intentus, -a, -um, adj. *keen, intent*
inter, prep. with acc. *among, between*
interaestuans, -ntis, part. *inflamed*
interdiū, adv. *in the day time*
interdum, adv. *sometimes*
intereā, adv. *meanwhile*
interficio, -ficere, -fēci, -fectum, tr. *kill, put to death*
interim, adv. *meanwhile*
interior, -ius, comp. adj. *inner*

intermitto, -ere, -mīsi, -missum, tr. *interrupt*
interpretor, 1, dep. tr. *conclude, consider*
interrogo, 1, tr. *ask*
intersum, -esse, -fui, intr. with dat. *be present at*
intrā, prep. with acc. *inside, within*
intro, 1, tr. *enter*
intueor, -ēri, -tuitus, dep. tr. *look at*
inūsitātus, -a, -um, adj. *unusual*
inūtilis, -e, adj. *useless*
invādo, -ere, -vāsi, -vāsum, tr. *attack*
invalesco, -ere, -ui, —, intr. *become strong*
invalidus, -a, -um, adj. *weak*
inveho, -ere, -vexi, -vectum, tr. *carry into*; in pass. *sail*
invenio, -īre, -vēni, -ventum, tr. *find*
invictus, -a, -um, adj. *unconquerable*
invideo, -ēre, -vīdi, -vīsum, intr. *envy*
invidus, -a, -um, adj. *grudging*
invīsus, -a, -um, adj. *hated*
invīto, 1, tr. *invite*
iocus, -i, m. *jest*
ipse, ipsa, ipsum, pron. *himself, herself, itself*
īra, -ae, f. *anger*
is, **ea**, **id**, pron. *that, he, she, it*
Issa, -ae, f. name of a pet dog
iste, ista, istud, adj. *that*
ita, adv. *thus, so, in such a way*
Ītalicus, -a, -um, adj. *Italian*
Ītalis, -idis, f. adj. *Italian*
Ītalus, -a, -um, adj. *Italian*
iter, itineris, n. *road, march*
iterum, adv. *again, a second time*
iubeo, -ēre, iussi, iussum, tr. *order*
iūcundus, -a, -um, adj. *dear, pleasant*
iūdex, -dicis, m. *judge*
iūdicium, -i, n. *opinion, reflection*
iūgerum, -i, n. *acre*
iugum, -i, n. *hill, ridge*
Iūlēus, -a, -um, adj. *of Julius Caesar*
Iūlius, -i, m. a Roman name
Iūpiter, Iovis, m. the god *Jupiter*
iūs, iūris, n. *right*; **iūre**, *rightly*
Iuvenālis, -is, m. the poet *Juvenal*
iuvenis, -is, m. *a youth*
iuvo, -āre, iūvi, iūtum, tr. *help, please, delight*

Kal. = **Kalendae**, -ārum, f. pl. the first day of each month

Lābīcānus, -a, -um, *of the Labican Way*
lābor, -i, lapsus, dep. intr. *glide*
labor, -ōris, m. *toil, work*
labōriōsus, -a, -um, adj. *laborious, hard to keep up*
lacerna, -ae, f. *cloak*
Lachesis, -is, f. one of the Fates
lacrima, -ae, f. *tear*
lacus, -ūs, m. *lake, pool*
laetitia, -ae, f. *happiness, joy*
laetor, 1, dep. intr. *rejoice*
laetus, -a, -um, adj. *happy, rich*
laevus, -a, -um, adj. *left* (hand)
lagōna, -ae, f. *flagon*

lambo, -ere, -bi, -bitum, tr. *lick*
langueo, -ēre, —, —, intr. *be idle*
languidus, -a, -um, adj. *sickly, feeble*
lanius, -i, m. *butcher*
lapillus, -i, m. *stone, gem, pearl*
lapis, -idis, m. *stone, mile-stone*
lār, laris, m. *household god*; pl. *house*
Larcius, -i, m. see **Macedo**
lascīvia, -ae, f. *playfulness, wantonness*
lascīvus, -a, -um, adj. *playful, wanton*
lassus, -a, -um, adj. *weary*
lateo, -ēre, -ui, —, intr. *lie hid*
Latīnus, -i, m. an actor
lātitūdo, -inis, f. *breadth, width*
Latius, -a, -um, adj. *Latin*
latrōcinium, -i, n. *robbery*
latus, lateris, n. *side, flank*
lātus, -a, -um, adj. *broad*; adv. **lātē**, *far and wide*
laudo, 1, tr. *praise*
laus, laudis, f. *praise*
lavo, -āre, lāvi, lavātum, lōtum or lautum, tr. *wash, bathe*
laxo, 1, tr. *relax, open up*
lector, -ōris, m. *reader*
lectulus, -i, m. *couch*
lectus, -i, m. *couch, bed*
lēgātus, -i, m. *legate*
lego, -ere, lēgi, lectum, tr. *read*
lēnio, 4, tr. *lessen*
lēnis, -e, adj. *gentle*; adv. **lēniter**
lentus, -a, -um, adj. *slow*
leo, -ōnis, m. *lion*
Lēthaeus, -a, -um, adj. *of Lēthē*, a river of Hell
levis, -e, adj. *light, false*; adv. **leviter**
lēvis, -e, adj. *smooth, close-cropped*
libellus, -i, m. *book, little book, accusation*
libens, -ntis, adj. *willing*; adv. **libenter**
liber, -bri, m. *book*
līberālis, -e, adj. *generous*
līberi, -ōrum, m. pl. *children*
lībertus, -i, m. *freedman*
libet, -ēre, -uit, impers. with dat. *it pleases*
lībum, -i, n. *cake*
Liburnica, -ae, f. *galley*
Libycus, -a, -um, adj. *Libyan*
Libys, -yos, acc. -yn, m. *a Libyan*
licet, -ēre, -uit, impers. with dat. *it is permitted*; with subjunctive, *although*; part. **licens**, *free, sportive*
lignum, -i, n. *wood*
ligo, -ōnis, m. *hoe*
līmen, līminis, n. *threshold, door*
līnea, -ae, f. *fishing-line*
lingua, -ae, f. *tongue, language*
lino, -ere, lēvi, litum, tr. *cover*
linteum, -i, n. *towel, napkin, canvas*
Līris, -is, m. river in Campania
līs, lītis, f. *law-suit*
littera, -ae, f. *letter*; pl. *letter, literature*
litūra, -ae, f. *erasure, correction*
lītus, -oris, n. *shore*
Līvius, -i, m. the historian *Livy*

locus, -i, m. *place, scene, opportunity, room*
longus, -a, -um, adj. *long, long-lasting, distant*; comp. adv. **longius**, *some way off*
loquor, -i, locūtus, dep. *speak*
lōtus, -a, -um, p.p.p. of lavo, *having bathed*
lucerna, -ae, f. *lamp*
Lucrīnus, -a, -um, adj. *of the Lucrine lake*, in Campania
luctus, -ūs, m. *grief*
lūdibrium, -i, n. *mockery, wanton attack*
lūdificor, 1, dep. *mock*
lūdo, -ere, lūsi, lūsum, intr. *play*
lūdus, -i, m. *play, school*
lūmen, lūminis, n. *light, eye*
lupa, -ae, f. *she-wolf*
Lupercus, -i, m. a mean friend of Martial's
Lupus, -i, m. a friend of Martial's
lupus, -i, m. *pike* (fish)
lūridus, -a, -um, adj. *pale*
lūsus, -ūs, m. *play, sportive verse*
lutum, -i, n. *mud*
lux, lūcis, f. *light, day, dawn*
Lyaeus, -i, m. *Bacchus*
lymphātus, -a, -um, p.p.p. *maddened*

Macedo, -ōnis, m. Larcius, an ex-praetor
maciēs, -iēi, f. *leanness*
maculōsus, -a, -um, adj. *mottled*
madeo, -ēre, -ui, —, intr. *be wet*
madidus, -a, -um, adj. *dripping, wine-soaked*
maestus, -a, -um, adj. *sad*
magis, comp. adv. *more*
magister, -tri, m. *master, captain, keeper*
magistrātus, -ūs, m. *magistrate*
magnitūdo, -dinis, f. *size*
magnus, -a, -um, adj. *great, powerful*
māior, māius, comp. adj. *larger, greater, broader*
mālo, malle, mālui, —, tr. *prefer*
malus, -a, -um, adj. *bad, wicked*; n. noun, malum, *misfortune*; adv. male, *badly, cruelly, scarcely, painfully*; **male dīcere**, *curse*
mando, 1, tr. *order, entrust*; mandātum, *order, message*
maneo, -ēre, mansi, -sum, intr. *remain, survive*
mānēs, -ium, m. pl. *shades, ghost*
manifestus, -a, -um, adj. *plain, obvious*
mansuētūdo, -dinis, f. *tameness*
manus, -ūs, f. *hand, band*
Marcus, -i, m. a Roman proper name, e.g. Martial's
mare, maris, n. *sea*
Marīca, -ae, f. a nymph living near the river Liris
marisca, -ae, f. *fig*
marītus, -i, m. *husband*
Maro, -ōnis, m. the poet *Vergil*
Mars, Martis, m. the god of war
Martiālis, -is, m. 1, the poet *Martial*; 2, one of his friends
Martius, -a, -um, adj. *of Mars, Roman*
māter, -tris, f. *mother, woman*

māteria, -ae, f. *theme, subject, story*
mātūtīnus, -a, -um, adj. *in the morning*
Maximus, -i, m. a friend of Pliny's
maximus, -a, -um, superl. adj. *largest, greatest*; adv. maximē, *especially, best*
meātus, -ūs, m. *course*; **meātus animae**, *breathing*
mediocriter, adv. *in a small degree*
medius, -a, -um, adj. *middle, intervening*
melior, -ius, comp. adj. *better*
memini, -isse, defect. intr. *remember*
memorābilis, -e, adj. *famous, memorable*
memoria, -ae, f. *memory, fame*
memoro, 1, tr. *describe, tell*
mens, mentis, f. *mind*
mensa, -ae, f. *table*
mensis, -is, m. *month*
mentior, 4, dep. intr. *tell lies*; p.p.p. **mentītus**, *feigned*
mergo, -ere, mersi, -sum, tr. *plunge*; in pass. *dive*
merito, adv. *deservedly*
merum, -i, n. *wine*
messis, -is, f. *harvest*
metallum, -i, n. *ore, metal*
metuo, -uere, -ui, -ūtum, tr. *fear*
metus, -ūs, m. *fear*
meus, -a, -um, adj. *my*
mīles, mīlitis, m. *soldier*
mille, indecl. *thousand*; pl. mīlia
Minerva, -ae, f. goddess of learning
minimus, -a, -um, superl. adj. *very small*
ministerium, -i, n. *service*
ministra, -ae, f. *deaconess*
minor, 1, dep. tr. *threaten*
minor, minus, comp. adj. *smaller, less, younger*; adv. **minus**, less
minuo, -uere, -ui, -ūtum, tr. *lessen, belittle*
mīrābilis, -e, adj. *marvellous, admirable*
mīrāculum, -i, n. *miracle, strange sight, strange story*
mīror, 1, dep. tr. *marvel, admire*; **mīrandus**, *marvellous*
mīrus, -a, -um, adj. *marvellous*
misceo, -ēre, -ui, mixtum, tr. *mingle, confuse*
Mīsēnum, -i, n. a port near Naples
miser, -era, -erum, adj. *wretched*
miserātio, -ōnis, f. *pity, pathos*
miseror, 1, dep. tr. *bewail*
mītis, -e, adj. *kind, soft*
mitto, -ere, mīsi, missum, tr. *send, throw*
modicus, -a, -um, adj. *small, modest*
modŏ, adv. *only, recently, just now, now...now*
modus, -i, m. *way, method*; **modo** or **in modum** with gen. *like*
moenia, -ium, n. pl. *town walls*
mōlēs, -is, f. *mass, pile*
molestus, -a, -um, adj. *troublesome*; **molestē ferre**, *be grieved*
mollis, -e, adj. *soft*
Molorchus, -i, n. a shepherd who entertained Hercules
moneo, 2, tr. *advise*
monimentum, -i, n. *memorial*
mons, montis, m. *mountain*

monstrum, -i, n. *horror*
mora, -ae, f. *delay*
morbus, -i, m. *disease*
morior, -i, mortuus, dep. intr. *die*; mortuus, *dead*
moror, 1, dep. intr. *delay*
mors, mortis, f. *death*
mortālitās, -ātis, f. *death, destruction*
mortiferē, adv. *fatally*
mōs, mōris, m. *custom, manner*; mōre with gen. *like*
mōtus, -ūs, m. *movement*
moveo, -ēre, mōvi, mōtum, tr. *move, take up*
mox, adv. *soon, then*
mūgil, -ilis, m. *mullet*
mulier, -eris, f. *woman*
mullus, -i, m. *barbel, mullet*
multitūdo, -inis, f. *number, crowd*
multus, -a, -um, adj. *much, many*; adv. multum, *much*
Mulvius (adj.) **pons**, *the Mulvian Bridge*
mundus, -i, m. *world*
mūniceps, -cipis, m. *fellow-townsman*
mūnīmentum, -i, n. *defence*
mūnus, mūneris, n. *gift*
mūraena, -ae, f. *lamprey*
murmur, -uris, n. *murmur, noise*
mūs, mūris, m. *mouse*
Mūsa, -ae, f. *Muse*
mustum, -i, n. *new wine*
mūto, 1, tr. *change*

nam, namque, conj. *for*
narro, 1, tr. *say, relate*
Nāso, -ōnis, m. the poet *Ovid*
nāsus, -i, m. *nose*
nātīvus, -a, -um, adj. *native, natural*
nato, 1, intr. and tr. *swim*
nātūra, -ae, f. *nature, quality*
nātus, -i, m. *son*
nauticus, -a, -um, adj. *of boatmen*
nāvicula, -ae, f. *little boat*
nāvigābilis, -e, adj. *fit for navigation*
nāvigium, -i, n. *ship*
nāvigo, 1, intr. *sail*
nāvis, -is, f. *ship*
nē, conj. *that...not, lest*, nē...quidem, *not even*
-ne, interrog. particle
nebula, -ae, f. *cloud*
nec, neque, conj. *and...not, nor, neither...nor*
necessārius, -a, -um, adj. *necessary*
nectar, -aris, n. *nectar*, the drink of the gods
nefās, indecl. n. *crime*
negligo, -ere, -lexi, -lectum, tr. *neglect*
nego, 1, tr. *deny, refuse, prevent*
negōtiōsus, -a, -um, adj. *busy*
nēmo, nēminem, m. and f. *nobody*; **nōn nēmo**, *several*
nempe, conj. *why, of course*
nemus, nemoris, n. *grove*
Nepōs, -ōtis, m. a friend of Pliny's
neptis, -is, f. *granddaughter*
nēquam, indecl. adj. *naughty, treacherous*
nēquāquam, adv. *by no means*
neque, nec, conj. *and...not, nor, neither...nor*
Nērēūs, -ei, m. a sea-god
nescio, 4, tr. *not to know*
Nestor, -oris, m. an aged Greek leader at Troy
neuter, -tra, -trum, adj. *neither*

nēve, conj. = **et nē**
nī, conj. *unless*
nīdulus, -i, m. *little nest*
nīdus, -i, m. *nest, pigeon-hole*
niger, -gra, -grum, adj. *black, dark*
nihil, nīl, indecl. n. *nothing*; abl. **nihilo**
nimium, adv. *too much*
nisi, conj. *unless, except*
niteo, -ēre, —, —, intr. *gleam*
nitidus, -a, -um, adj. *gleaming, splendid, oily*
nītor, -i, nīsus or nixus, dep. intr. *lean, rest*
nix, nivis, f. *snow*
no, 1, intr. *swim*
nocens, -ntis, adj. *guilty*
nocturnus, -a, -um, adj. *in the night*
nōdus, -i, m. *knot, girdle*
nōlo, nolle, nōlui, —, intr. *be unwilling*
nōmen, nōminis, n. *name, reputation*
nōmenculātor, -ōris, m. *announcer, footman*
nōmino, 1, tr. *name*
nōn, adv. *not*
nondum, adv. *not yet*
nonne, interrog. particle
nōnus, -a, -um, adj. *ninth*
nōs, nostrum or nostri, pron. *we, I*
noscito, 1, tr. *recognize*
nosco, -ere, nōvi, nōtum, tr. *get to know, inspect*
noster, -tra, -trum, adj. *our, my*
notābilis, -e, adj. *conspicuous, remarkable*
noto, 1, tr. *mark*
nōtus, -a, -um, adj. *well known, conspicuous*
novācula, -ae, f. *razor*
novitās, -ātis, f. *strangeness*
novus, -a, -um, adj. *new*; **novissimus**, *last*
nox, noctis, f. *night*
nūbēs, -is, f. *cloud*
nūbilus, -a, -um, adj. *cloudy*
nūdus, -a, -um, adj. *bare*
nullus, -a, -um, adj. *no, none*
nūmen, nūminis, n. *god, divinity*
numerus, -i, m. *number*
numquam, adv. *never*
nunc, adv. *now*
nuntio, 1, tr. *announce, declare*
nuntius, -i, m. *news*
nūper, adv. *lately*
nurus, -ūs, f. *daughter-in-law*
nusquam, adv. *nowhere*
nūto, 1, intr. *nod, totter*
nux, nucis, f. *nut*

ob, prep. with acc. *on account of*
obdūco, 3, tr. *cover*
obeo, -īre, -ii, -itum, tr. *discharge, carry out*
oblīdo, -ere, -līsi, -līsum, tr. *crush*
oblīvio, -ōnis, f. *forgetfulness*
obnoxius, -a, -um, adj. *exposed, open to*
obscūrus, -a, -um, adj. *obscure, secret, unknown*
obsideo, -ēre, -sēdi, -sessum, tr. *beset, throng*
obstinātio, -ōnis, f. *stubbornness*
obsto, -āre, -stiti, -stātum, intr. with dat. *hinder, block*
obstringo, -ere, -strinxi, -strictum, tr. *bind*

obstruo, -ere, -struxi, -structum, tr. *block up, choke*
obtero, -ere, -trīvi, -trītum, tr. *crush*
obtineo, -ēre, -ui, -tentum, tr. *hold, govern*
obversor, 1, dep. intr. with dat. *move about near*
occido, -ere, -cidi, -cāsum, intr. *die*
occīdo, -ere, -cīdi, -cīsum, tr. *kill*
occultē, adv. *secretly*
occupo, 1, tr. *seize, take up*
occurro, -ere, -curri, -cursum, intr. with dat. *meet, come to meet*
occurso, 1, intr. with dat. *keep meeting*
Octāvius, -i, m. see **Avītus**
octāvus, -a, -um, adj. *eighth*
oculus, -i, m. *eye*
odor, -ōris, m. *smell*
offero, -ferre, obtuli, oblātum, tr. *offer*; in pass. *appear*
officiōsus, -a, -um, adj. *doing one's duty*
officium, -i, n. *duty, service, social duty, attention, ceremony*
offirmo, 1, tr. *make firm*
oleo, -ēre, -ui, —, intr. *be scented*
ōlim, adv. *formerly*
olla, -ae, f. *pot*
ōminōsus, -a, -um, adj. *ill-omened*
omnis, -e, adj. *all, every, the whole*; adv. **omnīno**, *at all, ever*
opācus, -a, -um, adj. *shady*
operio, -īre, -ui, opertum, tr. *hide, cover, cover over*
opīnio, -ōnis, f. *opinion*
opīnor, 1, dep. tr. *infer*
oppidum, -i, n. *town*
oppleo, -ēre, -ēvi, -ētum, tr. *fill up*
opportūnē, adv. *appropriately*
opto, 1, tr. *desire, hope for*
opus, operis, n. *work*; **opus est**, *it is necessary*
ōra, -ae, f. *shore*
orbis, -is, m. *circle, world* (often with terrarum)
orbitās, -ātis, f. *bereaved state*
ordo, ordinis, m. *rank*
orior, -īri, ortus, dep. *rise up*
orno, 1, tr. *reward, adorn*
ōro, 1, tr. *beg*
ōs, ōris, n. *face, head, mouth, lips, utterance*
ŏs, ossis, n. *bone*
osculum, -i, n. *kiss, pet*
ostendo, -ere, -tendi, -tentum or -tensum, tr. *show*
ōtium, -i, n. *leisure, peace* (also pl.)

paedagōgium, -i, n. *pages' quarters*
paene, adv. *almost*
paenitentia, -ae, f. *change of mind, repentance*
paenitet, -ēre, -uit, impers. with acc. *repent*
Paetus, -i, m. Caecīna, husband of Arria
pāgina, -ae, f. *page* (of book)
palaestra, -ae, f. *wrestling-school*
Palātium, -i, n., also pl., *the Palatine Hill* at Rome, *Palace*
palleo, -ēre, -ui, —, intr. *be pale*
palma, -ae, f. *palm, hand*
palmes, -itis, m. *vine-shoot*

Pān, Pānos, acc. -a, m. a woodland god

pango, -ere, panxi, pepigi or pēgi, pactum or panctum, tr. *compose*

pār, paris, adj. *equal, well-matched*; adv. **pariter**, *together*

parco, -ere, peperci, parsum, intr. with dat. *spare*

parcus, -a, -um, adj. *sparing, unwilling*

parens, -ntis, c. *father, mother, parent, author*

pāreo, 2, intr. with dat. *obey*

pariēs, -ietis, m. *wall*

pario, parere, peperi, partum, tr. *earn, bear* (children)

Parius, -a, -um, adj. *Parian*, i.e. from Paros, an Aegean island

parma, -ae, f. *shield, gladiator*

paro, 1, tr. *prepare, provide, buy*; p.p.p. **parātus**, *ready*

pars, partis, f. *part, side, direction*; pl. *party, side*

parum, adv. *too little, not*

parvulus, -a, -um, adj. *tiny*

parvus, -a, -um, adj. *small, little*

pasco, -ere, pāvi, pastum, tr. *feed*

passer, -eris, m. *sparrow*

pastus, -ūs, m. *food*

pateo, -ēre, -ui, —, intr. *lie open, be open, be seen*

pater, patris, m. *father*

patientia, -ae, f. *endurance*

patior, -i, passus, dep. tr. *suffer, allow*

patria, -ae, f. *country, native land*

patrōnus, -i, m. *protector*

patruus, -i, m. *uncle*

patulus, -a, -um, adj. *capacious, gaping*

pauci, -ae, -a, pl. adj. *few*

pauculi, -ae, -a, pl. adj. *very few*

paulum, adv. *a little*

pavīmentum, -i, n. *floor*

pavor, -ōris, m. *panic*

pax, pācis, f. *peace, gentleness*

pecco, 1, intr. *make a mistake, sin*

pectus, -oris, n. *chest, breast*

pecūliāris, -e, adj. *all one's own*

pecūnia, -ae, f. *money*

pecus, pecoris, n. *flock*

pello, -ere, pepuli, pulsum, tr. *drive back, defeat*

peltātus, -a, -um, adj. *carrying a shield*

Penātēs, -ium, m. pl. *household gods, house*

pendeo, -ēre, pependi, —, intr. *hang, threaten*

pendulus,- a, -um, adj. *hanging, built on a hill*

pensio, -ōnis, f. *rent*

pensum, -i, n. *wool weighed out, thread*

per, prep. with acc. *through, by means of, round, during, in, for, along*

perago, -ere, -ēgi, -actum, tr. *accomplish, gather in*

percunctor, 1, dep. *enquire*

percutio, -cutere, -cussi, -cussum, tr. *strike*

perdo, -ere, -didi, -ditum, tr. *lose*

Perenna, -ae, f. see **Anna**

perennis, -e, adj. *everlasting*

pereo, -īre, -ii, -itum, intr. *perish, die*

perfero, -ferre, -tuli, -lātum, tr. *endure, carry out*

perfidus, -a, -um, adj. *faithless, false*
perfodio, -ere, -fōdi, -fossum, tr. *pierce*
perfruor, -i, -fructus, dep. intr. with abl. *enjoy*
pergo, -ere, perrexi, -rectum, intr. *proceed*
perīclitor, 1, dep. intr. *be in danger*
periculum, -i, n. *danger*
perimo, -ere, -ēmi, -emptum, tr. *kill*
periūrium, -i, n. *perjury*
perlego, -ere, -lēgi, -lectum, tr. *read through*
permaneo, -ēre, -mansi, -mansum, intr. *remain*
permitto, -ere, -mīsi, -missum, tr. *allow*
perneo, -ēre, -nēvi, -nētum, tr. *spin to the end*
perōdi, -isse, -ōsus, semi-dep. tr. *loathe*
perpetuitās, -ātis, f. *immortality*
perpetuus, -a, -um, adj. *everlasting*
perquam, adv. *very much*
persaepe, adv. *very often*
persequor, -i, -secūtus, dep. tr. *pursue, relate*
persevēro, 1, intr. *continue, persist*
persōna, -ae, f. *mask*
perterreo, 2, tr. *terrify*
pertinācia, -ae, f. *obstinacy*
pertrecto, 1, tr. *handle*
pervagor, 1, dep. tr. *spread through*
pervenio, -īre, -vēni, -ventum, intr. *arrive, reach*
pervigilo, 1, intr. *be awake all night*
pessimus, -a, -um, superl. adj. *worst, very bad*
pestilens, -ntis, adj. *haunted, unhealthy*
peto, -ere, -īvi or -ii, -ītum, tr. *ask, make for*
phantasma, -atis, n. *ghost*
phasēlus, -i, acc. -on, c. *boat*
philosophus, -i, m. *philosopher*
Phoebus, -i, m. the god of poetry
picātus, -a, -um, p.p.p. *pitched, sealed with pitch*
pictus, -a, -um, p.p.p. *painted*
piger, -gra, -grum, adj. *lazy*
pīla, -ae, f. *pillar*
pīnus, -i or -ūs, f. *pine-tree*
piper, -eris, n. *pepper*
Pirus, -i, f. *Pear-tree*, a place in Rome
piscātor, -ōris, m. *fisherman*
piscātōrius, -a, -um, adj. *used for fishing*
piscīna, -ae, f. *fish-pond*
piscis, -is, m. *fish*
piscor, 1, dep. intr. *fish*
pius, -a, -um, adj. *kindly, righteous*
placeo, 2, intr. with dat. *please*
placidus, -a, -um, adj. *peaceful, kindly*
plānus, -a, -um, adj. *level*; adv. plānē, *truly*
platanōn, -ōnis, acc. -ōna, m. *grove of plane-trees*
platanus, -i, f. *plane-tree*
Platea, -ae, f. a place in Spain
plērīque, -aeque, -aque, pl. adj. *most*
Plīnius, -i, m. *Pliny*
plūrimus, -a, -um, superl. adj. *very many, very much*

plūs, plūris, comp. adj. *more*; pl. *several*
pōculum, -i, n. *cup*
poēma, -atis, n. *poem*
Poenus, -a, -um, adj. *Carthaginian*; also m. noun
poēta, -ae, m. *poet*
poēticus, -a, -um, adj. *poetic*
pōmifer, -era, -erum, adj. *fruitful*
Pompēiānus, -a, -um, adj. *of Pompey*
Pompōniānus, -i, m. a friend of Pliny the Elder's
pōmum, -i, n. *apple*
pondus, -deris, n. *weight*
pōno, -ere, posui, positum, tr. *place, plant, lay aside, compare*
pontus, -i, m. *sea*
popīna, -ae, f. *cook-shop*
populor, 1, dep. tr. *devastate*
populus, -i, m. *people*
porrigo, -ere, -rexi, -rectum, tr. *stretch out, offer*
porticus, -ūs, f. *portico, porch*
portus, -ūs, m. *harbour*
posco, -ere, poposci, —, tr. *demand, ask for*
possideo, -ēre, -sēdi, -sessum, tr. *possess*
possum, posse, potui, —, intr. *be able*
post, prep. with acc. *after*
posteā, adv. *afterwards*
posterus, -a, -um, adj. *next*; m. pl. *posterity*
postis, -is, m. *door-post*
postquam, conj. *after, when*
postrēmo, superl. adv. *finally*
potens, -ntis, adj. *powerful*
potior, comp. adj. *better*; adv. **potius**, *rather*; superl. **potissimus**, *most important*
praebeo, 2, tr. *offer*
praecēdo, -ere, -cessi, -cessum, tr. and intr. *precede, go before, happen beforehand*
praecingo, -ere, -cinxi, -cinctum, tr. *gird, surround*
praeclārus, -a, -um, adj. *famous*
praeda, -ae, f. *plunder, spoil*
praedium, -i, n. *estate*
praedo, -ōnis, m. *robber*
praeeo, -īre, -ii, -itum, intr. *precede*
praefero, -ferre, -tuli, -lātum, tr. *prefer*
praemium, -i, n. *reward*
Praeneste, -is, f. and n. a town in Latium
praenuntius, -a, -um, adj. *heralding, foretelling*
praesens, -ntis, part. *present, in person*
praesertim, adv. *especially*
praesto, -āre, -stiti, -stitum or -stātum, tr. *provide, perform*
praesūmo, -ere, -sumpsi, -sumptum, tr. *take* or *perform beforehand*
praetendo, -ere, -tendi, -tentum, tr. *interpose*
praeter, prep. with acc. *besides*
praetereā, adv. *besides, moreover*
praetereo, -īre, -ii, -itum, tr. *pass by*; p.p.p. **praeteritus**, *past*; n. noun, *the past*
praetor, -ōris, m. *praetor, judge*
praetōrius, -a, -um, adj. *of praetorian rank, ex-praetor*
praevaleo, -ēre, -ui, —, intr. *prevail*
prandium, -i, n. *lunch*

prātum, -i, n. *meadow, grass*
prāvus, -a, -um, adj. *wicked, misplaced*
precor, 1, dep. tr. and intr. *pray, pray for*
premo, -ere, pressi, pressum, tr. *crush, press upon*
pretium, -i, n. *price*
Priāpus, -i, m. a garden god
prīmus, -a, -um, superl. adj. *first, front*; adv. **prīmum, prīmo**, *at first, first*
prior, prius, comp. adj. *first, former*; adv. **prius**, *first*
Priscus, -i, m. Cornēlius, a friend of Pliny's
pro, prep. with abl. *for, on behalf of*
probus, -a, -um, adj. *virtuous*
prōcēdo, -ere, -cessi, -cessum, intr. *come out, advance*
procella, -ae, f. *storm*
Procnē, -ēs, f. *a swallow*
prōconsul, -is, m. *pro-consul*
procul, adv. *far away*
Proculus, -i, m. a rich patron of Martial's
prōcurro, -ere, -curri and -cucurri, -cursum, intr. *run forward*
prōdeo (eo), intr. *come forward*
prōdigiōsus, -a, -um, adj. *monstrous*
prōdūco, 3, tr. *bring out*
prōfero, -ferre, -tuli, -lātum, tr. *put forward*
proficiscor, -i, -fectus, dep. intr. *start, depart*
profiteor, -ēri, -fessus, dep. tr. *confess, offer*
profundus, -a, -um, adj. *deep*; n. noun, *depth, pool*
prōicio, -icere, -iēci, -iectum, tr. *abandon*
proinde, adv. *therefore*
prōmiscuus, -a, -um, adj. *ordinary*
prōmitto, -ere, -mīsi, -missum, tr. *promise*; p.p.p. **prōmissus**, *long*
prope, adv. *near, nearly*
propero, 1, intr. *hasten*
propior, -ius, comp. adj. *nearer*
prōpōno, -ere, -posui, -positum, tr. *bestow, publish, put forward*
proprius, -a, -um, adj. *peculiar to, one's own*
propter, prep. with acc. *on account of, for*
prōripio, -ere, -ripui, -reptum, tr. *hurry forward*; **sē prōripere**, *rush out*
prōrumpo, -ere, -rūpi, -ruptum, intr. *burst forth*
proscrībo, -ere, -scripsi, -scriptum, tr. *advertise*
prōsequor, -i, -secūtus, dep. tr. *accompany, honour*
prospecto, 1, tr. *look at*
prosperus, -a, -um, adj. *successful, prosperous*
prōsum, prōdesse, prōfui, intr. with dat. *profit, avail*
prōtinus, adv. *at once*
prout, conj. *just as*
prōveho, -ere, -vexi, -vectum, tr. *carry forward*; in pass. *swim out*
prōvincia, -ae, f. *province*
prōvinciālis, -e, adj. *provincial*
proximus, -a, -um, superl. adj. *last, next, nearest, very near, close*
prūdens, -ntis, adj. *wise*
prūdentia, -ae, f. *wisdom*

publicus, -a, -um, adj. *public*; **rēs publica**, *the state*; **publicum**, *a public bath*; adv. **publicē**, *at the public expense*
Publius, -i, m. a friend of Martial's
pudet, -ēre, -uit, impers. with acc. *it shames*
pudor, -ōris, m. *shame*
puella, -ae, f. *girl, young wife*
puer, -eri, m. *boy, slave*
puerīlis, -e, adj. *boyish*
pugillārēs, -ium, m. pl. *writing-tablets*
pugio, -ōnis, m. *dagger*
pulcher, -chra, -chrum, adj. *handsome, glorious*
pulchritūdo, -inis, f. *beauty*
pulso, 1, tr. *strike, knock at, wash* (of waves)
pulverulentus, -a, -um, adj. *dusty*
pūmex, -icis, m. *pumice-stone*
pūnio, 4, tr. *punish*
purpura, -ae, f. *purple stripe, purple fan, purple*
pūrus, -a, -um, adj. *pure, bright*
puto, 1, tr. *think*
putrefactus, -a, -um, p.p.p. *putrefied*

quā, conj. *where, by which way*
quadrirēmis, -is, f. *quadrireme*
quaero, -ere, quaesīvi, -sītum, tr. *look for, ask, go to*
quālis, -e, adj. *of what sort, as*
quāliscunque, quālecunque, adj. *of whatever kind*
quam, adv. *how, as, than*
quamquam, conj. *although, and yet*
quamvīs, conj. and adv. *although, however much*
quando, adv. *when?*; after **sī**, *ever*
quantus, -a, -um, adj. *how great, how much, as much as*
quārē, adv. *why?*
quasi, adv. *as if, as though*
quasso, 1, tr. *shake*
quātenus, adv. *how far*
quatio, quatere, —, quassum, tr. *shake*
quattuor, indecl. adj. *four*
-que, enclitic conj. *and*
queror, -i, questus, dep. intr. *complain, whine*
quī, quae, quod, relative pron. *who, which*
quia, conj. *because*
quīcunque, quaecunque, quodcunque, relative pron. *whoever, whatever*; adv. **quōcunque**, *wherever*
quīdam, quaedam, quoddam or quiddam, pron. *a certain, some*
quidem, adv. *indeed*; **nē . . . quidem**, *not even*
quiēs, -ētis, f. *rest, quiet*
quiesco, -ere, quiēvi, -ētum, intr. *rest, sleep, stay*
quiētus, -a, -um, adj. *quiet*
quīlibet, quaelibet, quodlibet, indef. pron. *any one you like*; adv. **quamlibet**, *however much*,
quīn, **quīn immo**, **quīn etiam**, conj. *in fact, nay more*
quinque, indecl. adj. *five*
quintus, -a, -um, adj. *fifth*
quirītātus, -ūs, m. *crying*
quis, quid, interrog. pron. *who, what*; **quid**, *why?*
quis, qua, quid, indef. pron. after **sī**, *anyone, any*

quisquam, quaequam, quicquam, indef. pron. *anyone*
quisque, quaeque, quodque, indef. pron. *each*
quisquis, quidquid, relative pron. *whoever, whatever*
quīvīs, quaevīs, quodvīs or **quidvīs**, indef. pron. *anyone you like*
quō, adv. *whither*
quod, conj. *because*
quondam, adv. *once, formerly*
quoque, conj. *also*
quot, indecl. adj. *how many*
quotiens, adv. *whenever*

radiātus, -a, -um, p.p.p. *adorned with rays, rayed*
rādo, -ere, rāsi, rāsum, tr. *scrape, smoothe*
rāmus, -i, m. *branch*
rapidus, -a, -um, adj. *swift*
rapio, rapere, rapui, raptum, tr. *seize, snatch away*
rārus, -a, -um, adj. *few, seldom, fine*
rastrum, -i, n. *rake*
ratio, -ōnis, f. *reason, consideration*
ratis, -is, f. *ship*
raucus, -a, -um, adj. *hoarse*
recēdo, -ere, -cessi, -cessum, intr. *retreat, depart*
recens, -ntis, adj. *fresh*
recessus, -ūs, m. *retreat*
recīdo, -ere, -cīdi, -cīsum, tr. *cut off*
recipio, -cipere, -cēpi, -ceptum, tr. *receive*
recreo, 1, tr. *refresh*
Rectīna, -ae, f. a friend of the elder Pliny's
rectus, -a, -um, adj. *upright, straight*
recubo, -āre, —, —, intr. *lie down*
recumbo, -ere, -cubui, —, intr. *lie, recline, lie at table*
recūso, 1, tr. *refuse*
reddo, -ere, -didi, -ditum, tr. *return, restore, assign*
redeo, -īre, -ii, -itum, intr. *return*
redūco, 3, tr. *lead back*
redux, -ducis, adj. *returning*
refero, -ferre, rettuli, relātum, tr. *bring back, tell, relate, win, refer*; **grātiam referre**, *thank*
refugio, -fugere, -fūgi, —, intr. *flee back*; tr. *escape*
regno, 1, intr. *reign*
regnum, -i, n. *power, realm*
rego, 3, tr. *command, direct*
regredior, -i, -gressus, dep. intr. *return*
religio, -ōnis, f. *reverence*
relinquo, -ere, -līqui, -lictum, tr. *leave*
rēliquiae, -ārum, f. pl. *remains*
reliquus, -a, -um, adj. *remaining, the rest*
relūceo, -ēre, -luxi, —, intr. *gleam, gleam again*
relūcesco, -lūcescere, -luxi, —, intr. *grow light again*
remaneo, -ēre, -mansi, -mansum, intr. *remain*
remedium, -i, n. *remedy*
remitto, -ere, -mīsi, -missum, tr. *send back, put down, refer*; p.p.p. **remissus**, *relaxed, gentle*
renovo, 1, tr. *renew, smoothe over*
reparo, 1, tr. *renew*
repello, -ere, reppuli, repulsum, tr. *drive back*

repente, adv. *suddenly*
reperio, -īre, repperi, repertum, tr. *find*
repeto, -ere, -īvi and -ii, -ītum, tr. *return to, revive*
repōno, -ere, -posui, -positum, tr. *make up*
reprimo, -ere, -pressi, -pressum, tr. *check*
requīro, -ere, -quīsīvi, -quīsītum, tr. *ask for, search for, look for, ask*
rēs, rei, f. *event, thing, method, affair, fact, property*; **rēs publica**, *state, wealth*
resīdo, -ere, -sēdi, —, intr. *sit down, drop*
resorbeo, -ēre, —, —, tr. *suck back*; in pass. *ebb*
respicio, -spicere, -spexi, -spectum, tr. and intr. *look back, look back at*
respondeo, -ēre, -spondi, -sponsum, tr. *answer*
responsum, -i, n. *reply*
resulto, 1, intr. *rebound, echo*
retineo, -ēre, -ui, -tentum, tr. *hold, retain*
retro, adv. *back*
reus, -a, -um, adj. *on trial, accused*
reverenter, adv. *reverently*
reverto, -ere, -verti, -versum, tr. and intr. *turn back*; in pass. *return*
revoco, 1, tr. *recall*
revolvo, -ere, -volvi, -volūtum, tr. *ponder*; in pass. *plunge back*
rhīnocerōs, -ōtis, m. *rhinoceros*
Rhodius, -a, -um, adj. *of Rhodes*
rhombus, -i, m. *turbot*
rhonchus, -i, m. *sneer*
rīdeo, -ēre, rīsi, rīsum, tr. and intr. *laugh at, laugh, smile*
rigidus, -a, -um, adj. *hard, stern*
rīte, adv. *duly*
rōbustus, -a, -um, adj. *of oak, strong*
rogātor, -ōris, m. *beggar*
rogo, 1, tr. *ask, ask for*
Rōma, -ae, f. *Rome*
Rōmānus, -a, -um, adj. *Roman*
rosa, -ae, f. *rose*
roscidus, -a, -um, adj. *dewy*
rota, -ae, f. *wheel, carriage*
rubens, -ntis, part. *red*
Rubrae, -ārum, f. pl. *Saxa Rubra*, a town near Rome
Rūfīna, -ae, f. see **Claudia**
Rūfīnus, -i, m. Fadius, a friend of Pliny's
ūfus, -i, m. a Roman surname
ruīna, -ae, f. *collapse, fragment, ruinous clashing*
rumpo, -ere, rūpi, ruptum, tr. *break, interrupt*
ruo, -ere, rui, rutum, fut. part. ruitūrus, intr. *fall, collapse*
rursus, adv. *again*
rūs, rūris, n. *country, country estate*
rusticus, -a, -um, adj. *rustic, of the countryside*
rūta, -ae, f. *rue*

sacer, -cra, -crum, adj. *sacred*
sacrāmentum, -i, n. *oath*
sacrilegus, -a, -um, adj. *sacrilegious*
saeculum, -i, n. *age, generation, time*
saepe, adv. *often*
saeta, -ae, f. *hair, fishing-line*
saevus, -a, -um, adj. *savage, cruel*

sagum, -i, n. *cloak, blanket*
sal, salis, m. *salt, wit*
Salāria (via), -ae, f. *the Salarian* or *Salt Road* from Rome
salictum, -i, n. *willow-bed*
Salmacis, -cidis, f., a spring in the Lucrine lake
Salo, -ōnis, m. a river in Spain
salūber, -bris, -bre, adj. *healthy, moderate*
salūs, -ūtis, f. *safety*
salūtātor, -ōris, m. *morning caller*
salūto, 1, tr. *greet, salute, pay calls on*
salvus, -a, -um, adj. *safe*
sanctus, -a, -um, adj. *sacred, righteous*
sanguineus, -a, -um, adj. *bloody*
sapiens, -ntis, adj. *wise*
sapio, sapere, -īvi or -ii, —, *be wise, be critical, be right*
sarcina, -ae, f. *pack, burden*; pl. *baggage*
Sarmaticus, -a, -um, adj. *Sarmatian* (in eastern Europe)
satio, 1, tr. *sate, satisfy*
satis, adv. *enough, quite*; **satis facere**, *satisfy*
saxum, -i, n. *stone*
scālae, -ārum, f. pl. *stairs*
scelerātus, -a, -um, adj. *wicked*
scelus, -eris, n. *crime*
scientia, -ae, f. *knowledge*
scīlicet, adv. *in fact, of course*
scindo, -ere, scīdi, scissum, tr. *split*
scio, 4, tr. *know*
scrībo, -ere, scripsi, scriptum, tr. *write*
Scrībōniānus, -i, m. leader of a revolt against Claudius
scrīnium, -i, n. *book-case, desk*
scriptum, -i, n. *writing*
Scythicus, -a, -um, adj. *Scythian*
sē, sui, reflexive personal pron. *himself, herself, itself, themselves*
sēcēdo, -ere, -cessi, -cessum, intr. *depart*
sēcessus, -ūs, m. *retreat*
sēcrētus, -a, -um, adj. *secret*; n. noun, *seclusion*
secundum, prep. with acc. *according to*
secundus, -a, -um, adj. *second, favourable*
Secundus, -i, m. one of Pliny's names
sēcūritās, -ātis, f. *carelessness, freedom from care*
sēcūrus, -a, -um, adj. *safe, free from care, fearless*
sed, conj. *but*
sedeo, -ēre, sēdi, sessum, intr. *sit, be placed*
sēdēs, -is, f. *seat, foundation*
seges, -etis, f. *corn crop*
segnis, -e, adj. *idle*; adv. **segniter**, *idly*
semel, adv. *once*
sēmita, -ae, f. *path*
semper, adv. *always*
senātus, -ūs, m. *senate*
senectūs, -ūtis, f. *old age*
senesco, -ere, senui, —, intr. *grow old, die away*
senex, senis, m. *old man*; also adj. *old*
sensim, adv. *gradually*
sentio, -īre, sensi, sensum, tr. *feel, perceive, hear*
sepelio, -īre, -īvi, sepultum, tr. *bury*
sēpōno, -ere, -posui, -positum, tr. *lay aside*

septem, indecl. adj. *seven*
September, -bris, -bre, adj. *of September*
septimus, -a, -um, adj. *seventh*
sepulcrum, -i, n. *tomb*
sequor, -i, secūtus, dep. tr. *follow*
serēnus, -a, -um, adj. *calm*
sermo, -ōnis, m. *speech, talk, conversation*
serpens, -ntis, c. *snake*
serpo, -ere, serpsi, serptum, intr. *spread*
serta, -ōrum, n. pl. *garlands*
sērus, -a, -um, adj. *late*
servio, 4, intr. with dat. *serve, be a slave*
servo, 1, tr. *keep, keep to*
servulus, -i, m. *young slave, slave*
servus, -i, m. *slave*
Sētia, -ae, f. a town in Latium
seu, sīve, conj. *whether...or, or if*
sevēritās, -ātis, f. *sternness*
sevērus, -a, -um, adj. *stern*
sex, indecl. adj. *six*
sextus, -a, -um, adj. *sixth*
Sextus, -i, m. a Roman name
sexus, -ūs, m. *sex*
sī, conj. *if*
sīc, adv. *so, thus*
sicco, 1, tr. *dry*
siccus, -a, -um, adj. *dry*
sīcut, adv. *just as*
sīdus, sīderis, n. *star*
significo, 1, intr. *make a sign*
signum, -i, n. *sign*
silentium, -i, n. (also pl.) *silence*
Sīlius, -i, m. Ītalicus, the poet
silva, -ae, f. *wood*
similis, -e, adj. *like, similar*
similitūdo, -inis, f. *likeness*
simplex, -plicis, adj. *simple, harmless*
simplicitās, -ātis, f. *simplicity, frankness*
simul, adv. *at the same time*
simulācrum, -i, n. *likeness, image, phantom*
simulo, 1, tr. *pretend*
sine, prep. with abl. *without*
singuli, -ae, -a, pl. adj. *one by one, as individuals*
sinus, -ūs, m. *bay, bosom*
sisto, -ere, stiti, statum, tr. and intr. *stop*
situs, -a, -um, p.p.p. *situated*; **hīc situs est**, *here lies*
sīve, seu, conj. *whether...or, or if*
socer, soceri, m. *father-in-law*
socrus, -ūs, f. *mother-in-law*
sodālis, -is, m. *friend*
sōl, sōlis, m. *sun*
sōlācium, -i, n. *comfort*
solea, -ae, f. *slipper*
soleo, -ēre, -itus, semi-dep. intr. *be accustomed, be used*
sōlitūdo, -inis, f. *loneliness, abandoned district*
sollemnis, -e, adj. *customary*; n. noun, *a rite*
sollicito, 1, tr. *attract, rouse*
sōlor, 1, dep. tr. *relieve*
sōlus, -a, -um, adj. *alone*; adv. **sōlum**, *only*
solvo, -ere, solvi, solūtum, tr. *free, overcome*
somnus, -i, m. *sleep*
sono, -āre, -ui, —, intr. *sound, make a noise*; part. **sonans**, *loud*
sonus, -i, m. *sound*
sophōs, adv. *well done, bravo*
sordeo, -ēre, —, —, intr. *be disgraceful*

sordidus, -a, -um, adj. *dirty*
soror, -ōris, f. *sister*
spargo, -ere, sparsi, sparsum, tr. *scatter*
spatior, 1, dep. intr. *walk*
spatiōsus, -a, -um, adj. *large*
spatium, -i, n. *space, height, length*
speciēs, -iēi, f. *beauty, kind, appearance*
speciōsus, -a, -um, adj. *honourable*
spectāculum, -i, n. *sight*
spectātor, -ōris, m. *onlooker*
specto, 1, tr. *look at, see*
spēro, 1, tr. *hope*
spēs, spei, f. *hope*
spīritus, -ūs, m. *breath, wind, blast*
squālor, -ōris, m. *filth*
Stabiae, -ārum, f. pl. a town near Pompeii
stagnum, -i, n. *lake*
statim, adv. *immediately*
status, -a, -um, p.p.p. *appointed*
sterno, -ere, strāvi, strātum, tr. *overthrow, strew*
stilus, -i, m. *pen*
sto, stāre, steti, statum, intr. *stand*
stomachus, -i, m. *wind-pipe*
strepitus, -ūs, m. *noise, clatter*
stringo, -ere, strinxi, strictum, tr. *draw* (sword), *ruffle*
struēs, -is, f. *pile*
struo, -ere, struxi, structum, tr. *build, plan*
studeo, -ēre, -ui, —, intr. *study*
studiōsus, -a, -um, adj. *studious, scientific*
studium, -i, n. *pursuit, study, literary work*
Stygius, -a, -um, adj. *Stygian, hellish*
Stymphālides, -idum, f. pl. *the Stymphalian birds*
sūādeo, -ēre, sūāsi, sūāsum, intr. *advise*
sub, prep. with abl. *under, near*
subeo, -īre, -ii, -itum, tr. and intr. *approach, seize, come into one's mind, go under, take on one's back*
subiaceo, -ēre, -ui, —, intr. *lie below*
subinde, adv. *immediately*
subitus, -a, -um, adj. *sudden*; adv. subito, *suddenly*
subiungo, -ere, -iunxi, -iunctum, tr. *add*
subsisto, -ere, -stiti, —, intr. *halt, remain*
Subūra, -ae, f. *the Subura*, a district in Rome
succurro, -ere, -curri, -cursum, intr. *occur to one*
sūdātrix, -trīcis, f. adj. *causing sweat, hot, stuffy*
sūdo, 1, intr. *sweat*
sufficio, sufficere, -fēci, -fectum, intr. *satisfy, suffice*
sulpur, -uris, n. *sulphur*
sum, esse, fui, intr. *be*
summa, -ae, f. *total*
summitto, -ere, -mīsi, -missum, tr. *send up, let grow*
summus, -a, -um, superl. adj. *highest, top, greatest, last*
sūmo, -ere, sumpsi, sumptum, tr. *take*
sumptus, -ūs, m. *expense*
super, prep. with acc. *over, above, at*
superbus, -a, -um, adj. *proud*
supercilium, -i, n. *frown*

superfundo, -ere, -fūdi, -fūsum, tr. *pour in, pour on*
superi, -ōrum, m. pl. *gods*
superstes, -stitis, adj. *surviving*
superstitio, -ōnis, f. *superstition*
supersum, -esse, -fui, intr. *survive, remain over*
supplicium, -i, n. *punishment*
supplico, 1, tr. *worship*
suprēmus, -a, -um, superl. adj. *last*
Sūra, -ae, m. a friend of Pliny's
surgo, -ere, surrexi, -rectum, intr. *rise*
sūs, suis, m. *boar*
suscipio, -cipere, -cēpi, -ceptum, tr. *undertake*
suspectus, -a, -um, p.p.p. *suspected*
suspensus, -a, -um, p.p.p. *in suspense*
suspīrium, -i, n. *sigh, breathing*
suus, -a, -um, adj. *his own, her own, their own*
Syria, -ae, f. a Roman province

tabella, -ae, f. *plank, picture*
taberna, -ae, f. *shop*
taceo, 2, tr. and intr. *be silent about, be silent*
Tacitus, -i, m. Cornēlius, the historian
tacitus, -a, -um, adj. *silent*
talpa, -ae, c. *mole*
tālus, -i, m. *knuckle-bone*
tam, adv. *so*
tamen, adv. *however*
tametsi, conj. *although, and yet*
tamquam, adv. *as if, as though, as*
tandem, adv. *at last*
tango, -ere, tetigi, tactum, tr. *touch*
tantum, adv. *alone, only*
tantus, -a, -um, adj. *so great, so much*
Tarpēius, -a, -um, adj. *Tarpeian*
Tarraco, -ōnis, f. a town in Spain
Tartareus, -a, -um, adj. *hellish*
Tartessiacus, -a, -um, adj. *of Tartessus*, in Spain
Tascus, -i, m. a friend of the elder Pliny's
Tecta (via), -ae, f. *the Covered Way*, in Rome
tectum, -i, n. *roof, building*
tego, -ere, texi, tectum, tr. *cover, hide*
temerārius, -a, -um, adj. *bold*
tempero, 1, tr. *regulate*; p.p.p. **temperātus**, *temperate*
templum, -i, n. *temple*
tempto, 1, tr. *try, explore*
tempus, -oris, n. *time*
tendo, -ere, tetendi, tentum or tensum, intr. *make one's way*
tenebrae, -ārum, f. pl. *darkness*
teneo, -ēre, -ui, tentum, tr. *hold, remember, restrain, dwell in, fill, seize*
tener, -era, -erum, adj. *gentle, tender, young*
tenuis, -e, adj. *thin, poor, narrow, elegant*
tenuo, 1, tr. *narrow, thin out*
ter, adv. *thrice*
tergum, -i, n. *back, rear*
tero, -ere, trīvi, trītum, tr. *rub, tread*
terra, -ae, f. *earth, land*
terreo, 2, tr. *terrify*
terribilis, -e, adj. *terrible*

terrificus, -a, -um, adj. *dreadful*
terror, -ōris, m. *terror*
tertius, -a, -um, adj. *third*; adv. **tertio**, *a third time*
tetricus, -a, -um, adj. *stern*
Thalīa, -ae, f. the Muse of comedy and epigrams
Thetis, -idis, f. a sea-nymph, also *the sea*
tholus, -i, m. *dome*
Thrasea, -ae, m. husband of the younger Arria
Thymelē, -ēs, f. a dancer
Tiberis, -is, acc. -im, m. the river *Tiber*
Tībur, -uris, n. a town in Latium
timeo, 2, tr. and intr. *fear*
timor, -ōris, m. *fear*
titulus, -i, m. *inscription, advertisement*
Titus, -i, m. a Roman name
toga, -ae, f. *toga*
tollo, -ere, sustuli, sublātum, tr. *lift up, carry off, open*
tondeo, -ēre, totondi, tonsum, tr. *shear, cut* (the hair)
tono, -āre, -ui, —, intr. *thunder*
tonsor, -ōris, m. *barber*
tormentum, -i, n. *torture*
torrens, -ntis, m. *a torrent*
tortus, -a, -um, p.p.p. *twisted*
torus, -i, m. *cushion, couch*
torvus, -a, -um, adj. *stern*
tot, **totidem**, indecl. adj. *so many*
totiens, adv. *so many times*
tōtus, -a, -um, adj. *the whole, all*
tractātus, -ūs, m. *course of an enquiry*
trādo, -ere, -didi, -ditum, tr. *hand over, relate*
traho, -ere, traxi, tractum, tr. *draw, bring, take*
Traiānus, -i, m. the emperor *Trajan*
transeo, -īre, -ii, -itum, tr. *pass by*
transitus, -ūs, m. *way past*
tremens, -ntis, part. *trembling, quivering*
tremor, -ōris, m. *tremor, earthquake*
trepidātio, -ōnis, f. *panic, fear*
trepido, 1, intr. *be afraid*
trēs, **tria**, pl. adj. *three*
trīcēsimus, -a, -um, adj. *thirtieth*
triennium, -i, n. *period of three years*
trietēris, -idis, f. *period of three years*
trīgintā, indecl. adj. *thirty*
triplex, -licis, adj. *three-fold*
tristis, -e, adj. *sad, stern, dangerous*
tristitia, -ae, f. *sorrow, sternness*
triumphus, -i, m. *triumph*
tropa, adv. a gambling game; **lūdere tropa**, *to play 'tropa'*
truncus, -i, m. *trunk*
tū, **tui**, personal pron. *thou, you*
Tullius, -i, m. see **Cicero**
tum, adv. *then*
tumidus, -a, -um, adj. *swelling, hot-tempered*
tumor, -ōris, m. *swell, rise*
tumulus, -i, m. *mound, grave*
tunc, adv. *then*
tunica, -ae, f. *shirt*
turba, -ae, f. *crowd, throng*
tūs, **tūris**, n. *incense*
Tusculānus, **Tusculus**, -a, -um, adj. *of Tusculum*, a town near Rome
Tuscus, -a, -um, adj. *Etruscan*

tūtus, -a, -um, adj. *safe*
tuus, -a, -um, adj. *your*

ubi, adv. *where*
ubīque, adv. *everywhere*
ullus, -a, -um, adj. *any*
ulterior, -ius, comp. adj. *further*
ultio, -ōnis, f. *revenge*
ultrā, adv. *further, any more*
ululātus, -ūs, m. *wailing*
umbra, -ae, f. *shade*
unda, -ae, f. *wave*
unde, adv. *whence, from where*
unguentum, -i, n. *ointment*
unguis, -is, m. *nail, claw*
ūniversus, -a, -um, adj. *whole*; **in ūniversum**, *as a general rule*
ūnus, -a, -um, adj. *one*; adv. **ūnā**, *also*
Urbicus, -i, m. name of a Roman child
urbs, urbis, f. *city*
ursa, -ae, f. *she-bear*
ūrūca, -ae, f. *caterpillar*
ut, **uti**, adv. and conj. *as, when, since, so that, in order that*
utcunque, adv. *somehow or other*
uterque, utraque, utrumque, pron. *each of two, both*
ūtor, -i, ūsus, dep. intr. with abl. *use*
uxor, -ōris, f. *wife*

Vacerra, -ae, m. a Roman literary critic
vaco, 1, intr. *be open*
vādo, -ere, —, —, intr. *go*
vadum, -i, n. *ford, shallow*
vagor, 1, dep. *roam, walk about*
vagus, -a, -um, adj. *wandering, idly sporting*
valeo, 2, intr. *fare well, be powerful, be able*
Valerius, -i, m. one of Martial's names
vallis, -is, f. *valley*
vānesco, -ere, —, —, intr. *begin to vanish*
vānus, -a, -um, adj. *false, useless*
varius, -a, -um, adj. *varied*
vastus, -a, -um, adj. *huge, rough*
vātēs, -is, m. *poet*
vāticinātio, -ōnis, f. *prophecy*
-ve, enclitic conj. *or*
vehiculum, -i, n. *carriage*
vel, conj. *or, even, either...or*
vellus, -eris, n. *fleece*
vēlo, 1, tr. *cover*
velut, adv. *as though*
vēna, -ae, f. *channel, stream*
vēneo, -īre, -ii, -itum, intr. *be sold*
venerandus, -a, -um, adj. *hallowed, to be worshipped*
veneror, 1, dep. tr. *worship, respect*
venia, -ae, f. *pardon*
venio, -īre, vēni, ventum, intr. *come*
venter, -tris, m. *stomach*
ventilo, 1, tr. *fan*
ventus, -i, m. *wind*
verber, -eris, n. *lash, blow*
verbero, 1, tr. *strike*
verēcundia, -ae, f. *modesty, purity*
vereor, 2, dep. tr. *fear*
Vergilius, -i, m. the poet *Vergil*
Verginius, -i, m. Rūfus, Pliny's guardian and a famous general
verna, -ae, c. *a home-born slave*; also adj. *native*

vēro, adv. *indeed, but*
Vērōna, -ae, f. birthplace of the poet Catullus
versiculus, -i, m. *little verse*
versus, -ūs, m. *verse*
vertex, -icis, m. *head, summit*
verto, -ere, verti, versum, tr. *change, put to flight, overthrow*
vērum, adv. *but*
vērus, -a, -um, adj. *true, genuine*
Vesta, -ae, f. the goddess *Vesta*
vester, -tra, -trum, adj. *your*
vestīgium, -i, n. *track, spot*
vestio, 4, tr. *clothe*
vestis, -is, f. *clothes*
Vesuvius, -i, m. a volcano in Campania
veto, -āre, -ui, -itum, tr. *forbid, prevent*
vetus, -eris, adj. *old*
vexo, 1, tr. *trouble, throng*
via, -ae, f. *road, way*
viāticum, -i, n. *parting gift, travelling expenses*
vibrātus,-a,-um, adj. *quivering*
vicem, -is, f. *change*; **vice alternā** or **in vicem**, *in turn, for one's own part*
vīcīnus, -a, -um, adj. *near*; m. noun, *neighbour*
victima, -ae, f. *victim*
victor, -ōris, m. *victor*
vīcus, -i, m. *street, district*
video, -ēre, vīdi, vīsum, tr. *see, take care*; in pass. *seem*
vigilia, -ae, f. *wakefulness*
vigilo, 1, intr. *keep awake*; tr. *spend in wakefulness*
vīginti, indecl. adj. *twenty*
vīlica, -ae, f. *bailiff's wife*
vīlicus, -i, m. *bailiff*
vīlis, -e, adj. *worthless*
vīlitās, -ātis, f. *cheapness*
villa, -ae, f. *country house*
vinco, -ere, vīci, victum, tr. and intr. *overcome, be victorious*
vinculum, -i, n. *fetter*
Vindex, -icis, m. Julius, a rebel against Nero
vindico, 1, tr. *avenge*
vīnum, -i, n. *wine*
viola, -ae, f. *violet*
vīpera, -ae, f. *snake*
vir, viri, m. *man*
vireo, -ēre, -ui, —, intr. *be green*
virga, -ae, f. *twig, shoot*
virgineus, -a, -um, adj. *of a virgin*
virgo, -inis, f. *girl*
virtūs, -ūtis, f. *virtue, skill*
vīs, vim, f. (pl. vīrēs), *strength*
viscera, -um, n. pl. *heart, vitals*
vīta, -ae, f. *life*
vīvo, -ere, vixi, victum, intr. *live*
vīvus, -a, -um, adj. *living, alive*
vix, adv. *scarcely*
voco, 1, tr. *call, summon*
volito, 1, intr. *fly*
volo, 1, intr. *fly*
volo, velle, volui, tr. *wish*
volūmen, -inis, n. *roll, book*
voluptās, -ātis, f. *pleasure*
vox, vōcis, f. *voice, word, saying*
vulgus, -i, n. *common people, crowd*
vulnus, -eris, n. *wound*
vultus, -ūs, m. *face*